STATISTICS
AND TRUTH

Putting Chance to Work

second edition

C. Radhakrishna Rao

Pennsylvania State University, USA

STATISTICS
AND TRUTH

Putting Chance to Work

second edition

World Scientific
Singapore • New Jersey • London • Hong Kong

Published by

World Scientific Publishing Co. Pte. Ltd.

P O Box 128, Farrer Road, Singapore 912805

USA office: Suite 1B, 1060 Main Street, River Edge, NJ 07661

UK office: 57 Shelton Street, Covent Garden, London WC2H 9HE

Librarty of Congress Cataloging-in-Publication Data
Rao, C. Radhakrishna (Calyampudi Radhakrishna), 1920–
    Statistics and truth : putting chance to work / C. Radhakrishna
Rao. -- 2nd enl. ed.
      p.   cm.
    Includes bibliographical references and index.
    ISBN 9810231113 (alk. paper)
    1. Mathematical statistics.   I. Title
QA276.16.R36   1997
519.5--dc21
                                           97-10349
                                             CIP

British Library Cataloguing-in-Publication Data
A catalogue record for this book is available from the British Library.

First published 1997
Reprinted 1999

Printed in Singapore.

Knowledge is what we know
Also, what we know we do not know.
We discover what we do not know
Essentially by what we know.

Thus knowledge expands.

With more knowledge we come to know
More of what we do not know.

Thus knowledge expands endlessly.

* * *

All knowledge is, in final analysis, history.
All sciences are, in the abstract, mathematics.
All judgements are, in their rationale, statistics.

Foreword

Beginning this year, CSIR has instituted a distinguished lectureship series. The objective here is to invite eminent scientists from India and abroad for delivering a series of three lectures on topics of their choice. The lectures, known as the CSIR Distinguished Lectures, were to be delivered in different locations of the country. The first of this series has been dedicated to the memory of the mathematics genius Srinivasa Ramanujan.

It augurs well that this first set of lectures (CSIR Ramanujan lectures) has begun with those of Prof. C. Radhakrishna Rao, National Professor (and currently Eberly Professor of Statistics, Penn State University), a distinguished scientist in the international statistics scene.

The lectures were delivered at the National Physical Laboratory in Delhi, at the Central Leather Research Institute in Madras and in the Indian Statistical Institute at Calcutta and were widely appreciated by professional statisticians, by physicists, chemists and biological scientists, by students of different age groups and by professionals and administrators. The scope of these lectures was wide and pervaded many areas of human activities, both scientific and administrative.

By arranging to have the lectures published now, CSIR hopes that a wider community of scientists the world over will be able to derive the benefit of the expertise of renowned men like Prof. Rao.

I express my appreciation of the efforts of Dr. Y.R. Sarma for having edited and brought out the volume quickly.

New Delhi,
December 31, 1987

A.P. MITRA
Director-General
Council of Scientific &
Industrial Research

*Uncertain
knowledge*

+

*Knowledge of the
amount of
uncertainty in it*

=

*Usable
knowledge*

Preface

I consider it a great honour to be called upon to deliver the Ramanujan Memorial Lectures under the auspices of the CSIR (Council of Scientific & Industrial Research). I would like to thank Dr. A.P. Mitra, Director General of the CSIR, for giving me this honour and opportunity to participate in the Ramanujan centenary celebrations.

I gave three lectures, the first one in Delhi, the second in Calcutta and the third in Madras as scheduled, which I have written up in four chapters for publication. In the beginning of each lecture I have said a few words about the life and work of Ramanujan, the rare mathematical genius who was a legendary figure in my younger days. This is to draw the attention of the younger generation to the achievements of Ramanujan, and to emphasize the need to reform our educational system and reorganize our research institutes to encourage creativity and original thinking among the students.

When I was a student, statistics was in its infancy and I have closely watched its evolution over the last 50 years as an independent discipline of great importance and a powerful tool in acquiring knowledge in any field of enquiry. The reason for such phenomenal developments is not difficult to seek.

Statistics as a method of learning from experience and decision making under uncertainty must have been practiced from the beginning of mankind. But the inductive reasoning involved in these processes has never been codified due to the uncertain nature of the conclusions drawn from given data or information. The breakthrough occurred only in the beginning of the present century with the realization that inductive reasoning can be made precise by specifying the amount of uncertainty involved in the conclusions drawn. This paved the way for working out an optimum course of action involving minimum risk, in any given uncertain situation, by a purely deductive process. Once this mechanism was made available, the flood gates opened and there was no end to the number of applications impatiently awaiting methods which could really deliver the goods.

From the time of Aristotle to the middle of the 19th century, chance was considered by scientists as well as philosophers to be an indication of our ignorance which makes prediction impossible. It is now recognized that chance is inherent in all natural phenomena, and the only way of understanding nature and making optimum predictions (with minimum loss) is to study the laws (or the inner structure) of chance and formulate appropriate rules of action. Chance may appear as an obstructor and an irritant in our daily life, but chance can also help and create. We have now learnt to put chance to work for the benefit of mankind.

I have chosen to speak on the foundations, modern developments and future of statistics, because of my involvement with statistics over the last 45 years as a teacher, research worker and consultant in statistics, and as an administrator managing the academic affairs of a large organization devoted to statistics. I grew up in a period of intensive developments in the history of modern statistics.

As a student I specialized in mathematics - the logic of deducing consequences from given premises. Later I studied statistics - a rational approach to learning from experience and the logic of identifying the premises given the consequences. I have come to realize the importance of both in all human endeavours whether it is in the advancement of natural knowledge or in the efficient management of our daily chores. I believe:

> *All knowledge is, in final analysis, history.*
> *All sciences are, in the abstract, mathematics.*
> *All judgements are, in their rationale, statistics.*

The title of my Lectures, *Statistics and Truth*, and their general theme are somewhat similar to those of *Probability, Statistics and Truth*, the collected lectures of R. von Mises published several years ago. Since the latter book appeared, there have been new developments in our thinking and attitude towards chance. We have reconciled with the "Dice-Playing God" and learnt to plan our lives to keep in resonance with uncertainties around us. We have begun to understand and accept the beneficial role of chance in situations beyond our control or extremely complex to deal with. To emphasize this I have chosen the subtitle, *Putting Chance to Work.*

Dr. Joshi, the Director of the National Physical Laboratory, reminded me what Thomas Huxley is reported to have said, that a man of science past sixty does more harm than good. *Statistically* it may be so. As we grow old we tend to stick to our past ideas and try to propagate them. This may not be good for science. Science advances by change, by the introduction of new ideas. These can arise only from uninhibited young minds capable of conceiving what may appear to be impossible but which may contain a nucleus for revolutionary change. But I am trying to follow Lord Rayleigh who was an active scientist throughout his long life. At the age of sixty-seven (which is exactly my present age), when asked by his son, who is also a famous physicist, to comment on Huxley's remark, Rayleigh responded:

> That may be if he undertakes to criticize the work of younger men, but I do not see why it need be so if he sticks to things he is conversant with.

However J.B.S. Haldane used to say that Indian scientists are polite and they do not criticize the work of each other, which is not good for the progress of science.

It gives me great pleasure to thank Dr. Y.R.K. Sarma of the Indian Statistical Institute for the generous help he has given in editing the Ramanujan Memorial Lectures I gave at various places in the form of a book and looking after its publication.

Calcutta
December 12, 1987   C.R. Rao

He who accepts statistics indiscriminately

will often be duped unnecessarily. But

he who distrusts statistics indiscriminately

will often be ignorant unnecessarily.

Preface to the Second Edition

The first edition of **Statistics and Truth: Putting Chance to Work** was based on three lectures on the history and development of statistics I gave during Ramanujan Centenary Celebrations in 1987. The topics covered in each lecture were reproduced in a more detailed form as separate sections of the book. The present edition differs from the first in many respects.

The material which appears under lectures 1, 2 and 3 in the first edition is completely reorganized and expanded to a sequence of five chapters to provide a coherent account of the development of statistics from its origin as collection and compilation of data for administrative purposes to a fullfledged separate scientific discipline of study and research. The relevance of statistics in all scientific investigations and decision making is demonstrated through a number of examples. Finally a completely new chapter (sixth) on the public understanding of statistics, which is of general interest, is added.

Chapter 1 deals with the concepts of randomness, chaos and chance, all of which play an important role in investigating and explaining natural phenomena. The role of random numbers in confidential transactions, in generating unbiased information and in solving problems involving complex computations is emphasized. Some thoughts are expressed on creativity in arts and science. Chapter 2 introduces the deductive and inductive methods used in the creation of new knowledge. It also demonstrates how quantification of uncertainty has led to optimum decision making,

Statistics has a long antiquity but a short history. Chapters 3 and 4 trace the development of statistics from the notches made by the primitive man to keep an account of his cattle to a powerful logical tool for extracting information from numbers or given data and drawing inferences under uncertainty. The need to have clean data free from faking, contamination or editing of any sort is stressed, and some methods are described to detect such defects in data.

Chapter 5 deals with the ubiquity of statistics as an inevitable tool in search of truth in any investigation whether it is for unravelling the mysteries of nature or for taking optimum decisions in daily life or for settling disputes in courts of law.

We are all living in an information age, and much of the information is transmitted in a quantitative form such as the following. The crime rate this year has gone down by 10% compared to last year. There is 30% chance of rain tomorrow. The Dow Jones index of stock market prices has gained 50 points. Every fourth child born is Chinese. The percentage of people approving President's foreign policy is 57 with a margin of error of 4 percentage points. You lose 8 years of your life if you remain unmarried. What do these numbers mean to the general public? What information is there in these numbers to help individuals in making right decisions to improve the quality of their lives. An attempt is made in Chapter 6 of the new edition to emphasize the need for public understanding of statistics, and what we can learn from numbers to be efficient citizens, as emphasized by H.G. Wells:

> Statistical thinking will one day be as necessary for efficient citizenship as the ability to read and write.

At the beginning of each lecture delivered in 1987, some aspects of Ramanujan's life and work were mentioned. All these biographical details are put together as a connected essay on Ramanujan in the Appendix to the book.

University Park,   C.R. Rao
March 31, 1997

Contents

Science is built up with fact as a house

is with stone. But a collection of facts is no more

a science than a heap of stones is a house.

Jules Henri Poincare

Chapter 1

Uncertainty, Randomness and Creation of New Knowledge

> *Let chaos storm!*
> *Let cloud shapes swarm!*
> *I wait for form.*
>
> — Robert Frost, *Pertinax*

1. Uncertainty and its quantification

The concepts of uncertainty and randomness have baffled mankind for a long time. We face uncertainties every moment in the physical and social environment in which we live. Chance occurrences hit us from every side. We bear with uncertainties of nature and we suffer from catastrophes. Things are not deterministic as Gothe wished:

> Great, eternal, unchangeable laws prescribed the paths along which we all wander?

or as Einstein, the greatest physicist in three centuries or possibly of all time, believed:

> God does not play dice with the universe.

Some theologians argue that nothing is random to God because he causes all to happen; others say that even God is at the mercy of some random events. In his book *"The Garden of Epicurus"* Anatole France remarks:

> Chance is perhaps the pseudonym of God when he did not want to sign.

1

Philosophers from the time of Aristotle acknowledged the role of chance in life, attributed it to something which violates order and remains beyond one's scope of comprehension. They did not recognize the possibility of studying chance or measuring uncertainty. The Indian philosophers found no need to think about chance as they believed in the ancient Indian teaching of Karma which is a rigid system of cause and effect explaining man's fate through his actions in previous lives.

All human activity is based on forecasting, whether entering a college, taking a job, marrying or investing money. Since the future is unpredictable however much information we have, there may not be any system of correct decision making. Uncertain situations and the inevitable fallibility in decision making have led mankind to depend on pseudosciences like astrology for answers, seek the advise of soothsayers or become victims of superstition and witchcraft. We still seem to rely on old wisdom:

> This is plain truth: every one ought to keep a sharp eye on main chance.
> - Plautus (200 B.C.)

This is still echoed in present day statements like:

> A chance may win that by mischance was lost.
> - Robert Southwell (1980)

Our success or failures are explained more in terms of chance than by our abilities and endeavors.

Uncertainty in any given situation can arise in many ways. It may be due to

* lack of information
* unknown inaccuracies in available information
* lack of technology to acquire needed information
* impossibility of making essential measurements
* ...

Uncertainty is also inherent in nature as in the behavior of fundamental particles in physics, genes and chromosomes in biology and individuals in a society under stress and strain, which necessitates the development of theories based on stochastic rather than deterministic laws in natural, biological and social sciences.

How do we take decisions under uncertainty? How do we generalize from particular observed data to discover a new phenomenon or postulate a new theory? Is the process involved an art, technology or science?

Attempts to answer these questions have begun only in the beginning of the present century by trying to quantify uncertainty. We may not have fully succeeded in this effort but whatever is achieved has brought about a revolution in all spheres of human endeavor. It has opened up new vistas of investigation and helped in the advancement of natural knowledge and human welfare. It has changed our way of thinking and enabled bold excursions to be made into the secrets of nature, which our inhibitions about determinism and inability to deal with chance prevented us earlier.

A full account of these developments and the reasons for the long delay in conceiving these ideas are given in the next chapter.

2. Randomness and random numbers

Strangely enough, the methodology for exploring uncertainty involves the use of randomly arranged numbers, like the sequence of numbers we get when we draw tokens numbered 0, 1, ..., 9 from a bag one by one, each time after replacing the token drawn and thoroughly shuffling the bag. Such sequences, called random numbers, are supposed to exhibit maximum uncertainty (chaos or entropy) in the sense that given the past sequence of digits drawn, there is no clue for predicting the outcome of the next draw. We shall see how they are generated and how indispensable they are in certain investigations and in solving problems involving complex computations.

2.1 A Book of Random Numbers!

In 1927, a statistician by the name L.H.C. Tippett produced a book titled Random Sampling Numbers. The contents of this book are 41,600 digits (from 0 to 9) arranged in sets of 4 in several columns and spread over 26 pages. It is said that the author took the figures of areas of parishes given in the British census returns, omitted the first two and last two digits in each figure of area and placed the truncated numbers one after the other in a somewhat mixed way till 41,600 digits were obtained. This book which is nothing but a haphazard collection of numbers became the best seller among technical books. This was followed by another publication by two great statisticians, R.A. Fisher and F. Yates, which contains 15,000 digits formed by listing the 15-19th digits in some 20 figure logarithm tables.

A book of random numbers! A meaningless and haphazard collection of numbers, neither fact nor fiction. Of what earthly use is it? Why are scientists interested in them? This would have been the reaction of the scientists and laymen in any earlier century. But a book of random numbers is typically a twentieth century invention arising out of the need for random numbers in solving problems of the real world. Now the production of random numbers is a multibillion dollar industry around the world involving considerable research and sophisticated high speed computers.

What is a sequence of random numbers? There is no simple definition except a vague one as mentioned earlier that it does not follow any particular pattern. How does one generate such an ideal sequence of numbers? For instance, you may toss a coin a number of times and record the sequence of 0's (for tails) and 1's (for heads) such as the following

$$011010\ldots\ldots$$

If you are not a magician who can exercise some control on each toss, you get a random sequence of what are called binary digits (0's and 1's). Such a sequence can also be obtained by drawing beads from a bag containing black and white beads in equal numbers, writing, say 0 for black bead drawn and 1 for white. When I was teaching the first year class at the Indian Statistical Institute, I used to send my students to the Bon-Hooghly Hospital near the Institute in Calcutta to get a record of successive male and female births delivered. Writing M for a male birth and F for a female birth we get a binary sequence as the one obtained above by repeatedly tossing a coin or drawing beads. One is a natural sequence of biological phenomena and another is an artificially generated one.

Table 1.1 gives a sequence of the colors of 1000 beads drawn with replacement from a bag containing equal numbers of white (W) and black (B) beads. Table 1.2 gives a sequence of 1000 Children delivered in a hospital according to sex of the child, male (M) or female (F). We can summarize the data of Tables 1.1 and 1.2 in the form of what are called frequency distributions. The frequencies of 0, 1, 2, 3, 4, 5 males in sets of 5 consecutive births and of white beads in sets of 5 consecutive draws of beads are given in Table 1.3.

The expected frequencies are theoretical values which are realizable on the average if the experiment with 200 trials is repeated a large number of times. The frequencies can be represented graphically in the form of what are called histograms.

It is seen that the two histograms are similar indicating that the chance mechanism of sex determination of a child is the same as that of drawing a black or a white bead from a bag containing equal numbers of beads of the two colors or similar to that of coin tossing. A simple exercise such as the above can provide the basis for formulating a theory of sex determination. God is tossing a coin! In fact statistical tests showed that the male-female births provide a more faithful random binary sequence than the artificially generated one. Perhaps God is throwing a more perfect coin. In India one child is born every second, which provides a cheap and expeditious source for generating binary random sequences.

Table 1.1 Data on color of successive beads drawn from a bag containing equal numbers of white and black heads

```
B W W B W    B W W B B    B B B W B    B B W W B    W W W B B
B W B B B    B B W W B    W B W W W    B B W W W    W W W W B
W W B W W    W B B W B    W W W B B    B B B W W    B W B W W
B W W W W    B B W B B    W W B B W    B W W B B    W B B W B
W B W B W    B W B B W    B B B B W    B B B B B    B B W B W
W B W B B    W B W B B    W B W B W    B W B B B    W W B B B
B W W B B    B W W B W    B W B B W    B W B B B    W B W B W
B B B W W    W W W B W    W B W W W    W W W B B    B B W W B
B B B W W    B W W W B    B B W W W    W W B B W    B B B W W
W W B B W    W W B W B    B B W B W    B W W W W    W B W B W

B W B B B    W W W B W    B W B B B    W B B W W    W B W B B
W B W B W    W W B W B    W W B W W    B W W W B    B B B W B
W W W W B    B B W W W    W W W W W    B B B B W    W W B W W
B W B W B    B B B W W    B W W W W    B W B B W    W B B B B
B B W B B    B B W W W    B W B W W    B W B W W    B B B W B
W W W B W    B W W W W    W W W W B    B B W B W    W W W B B
W W B W B    W W W B B    B B B W W    B W B W W    W W W B W
B B B W B    B W W W B    B W W B B    B B W B W    B B B B B
W W B W B    W B W W W    W B B B W    B B W B B    W B W W B
B W B W B    B B W B B    B B B B B    B B W B W    W W W W B

B W B W B    W W B B B    B B W W B    B W R W B    W W B B B
B W B W B    W W B B B    B B W W B    B W B W B    W W B B B
W W W B W    W B B B B    W W W W B    B W W W B    B B B B B
W B B W W    B B B W B    W W B B B    W W B W W    W W B B B
B B B B W    W B W B B    W W B W W    B B B W W    B W B W W
W W B W B    W B W B W    W B W W B    W B W B W    B B B W W
B W B W B    W W W W W    B W W W B    B B W B W    B W B W W
B B B B W    W B W W W    W W B B W    B W W W W    B B B W W
W B W B B    W B W W W    W W B W B    W W W B B    B B B W W
W B W B B    B B W B B    W B B W W    W B W B W    B W W B B

W B W W W    B B B B W    W B B B W    B W W W W    W B B W B
W B W B B    W B B W W    W W W W W    W B B W B    B B W W B
W B B W W    B B B B B    B W W B B    B W W W B    W B B W W
W W B B W    W W W B B    W W W B W    B B W B W    B W B B W
W B W B W    W B W B W    W B B B B    W B W W W    B W B B W
B W W B B    W B B B B    W W W W B    B W W W W    B W B W W
B W B W B    B W B B W    W B W B W    B W W W W    W B B W B
B B W B W    W B W B B    W W W B B    B W B B B    W B W B W
B B W W W    B W W B W    W W W B B    B B B W W    B W B W W
W W W B W    B B W B B    B W B B W    B W W W W    W W W W W
```

HISTOGRAM FOR TWO DATA SETS

WHICH IS WHICH ?

n = 200   n = 200

Table 1.2 Sequence of male (M) and female (F) children delivered in Bon-Hooghly Hospital, Calcutta

January
```
F M M F F   M M M M F   M F M F M   M M F F M   F F M F F
F M F M M   M M M M F   M M M M M   F F F F M   M F M M M
M M M M M   M M F M F   M M F F F   M M F M M   F F F M F
F M F M M   M F M M M   F F M M F   M F F M M   F M F M M
F F M F M   M F M F F   F M M F F   M F M F F   F M M M F
F F M F M   F M M M M   M F M F F   M F M F M   M F M M F
F F F F F   F F F M M   F M M M F   M M M M F   F M F F F
F M F M M   M M F F F   F M F F F   M M M M M
```

February
```
                                                F F M F F
F F M M M   F F F F M   F F F M F   F M F F M   F F M F F
M M M F M   H F M F M   F F M F M   M F M F M   M M F M M
F M M F F   F M M M F   F F F F M   M M F F F   M M F F M
M F M F M   F M M M M   F F M M F   F M M F M   F M M F M
F F
```

March
```
        M F F   F M M M M   M M M F M   F F F F F   M M M F M
M F M F F   M F M F F   F F F M M   F M F F M   F M M F M
M F F F F   F M M F M   F M M F F   M M M M M   M M F F M
M M F F M   M M M F M   F F M F M
```

April
```
                                                F M F F M   F F M M M
F F M F M   M F F F M   F M M F F   M F F F M   M F F M F
F M F M M   M M M F M   M M M M M   F F M M M   F M F M F
M M F M M   M M F F M   F M M M M   M M M F F   F M M F M
F M F F M   M F M F F   M M F M F   M F M F M   F F M F M
F F F F M   F M M M F   F M F F F   M M F F F   M M M F F
F F M F F   F M M M F   F M F M F   M F M F M   M M F M F
M F M M F   F M M F F   F M M F M   M M M M M   F M M F F
```

July
```
F M M M M   F M M M M   F F M F F   F F M M F   F M F M M
F F M M M   F M F F F   F M F M M   F M F M M   M M M M M
M F M F F   M M M M M   F M F M M   M F M M F   F M F M F
M F M M F   F F M M M   M M M F M   M M F F M   M M M F F
F M F F M   M F M F F   F F F F F   M M M M F   F F F M M
F F M M M   M M M M F   M M M M F   F M M F F   F F F M M
F
```

October
```
        M M M F   F F F M F   F M M F M   M F M M F   M M M M M
M F M F M   F F F F M   F M F F F   F M F M M   M F F F M
M F M M F   M M F F F   F F M F F   F M M M M   M F M M F
F M M F F   M F M M F   F F M F F   M M F F M   F F F M F
F M M F F   M M F M M   M M M M F   F M F F M   M F F M F
F F M M F   F F F M F   F F M F F   F F M F M   F F M F F
M M F M M   F F F M F   M F F M F   M M M F F   F F F F F
M F M F M   M M F M F   M F F M F   M M F M F   M M M F M
M F M M F   M M F F F   F F M F M   F F F M M   M F M M M
M F F F M   M F M F F   M F F F M   M M F F M   M F M M M
M F M M F   F M M M F   F F M M F   F F F F F   F F F M F
M M F M M   M F M F F
```

The survey was conducted by Srilekha Basu, a first year student. The data refer to births in some months in 1956.

Table 1.3 Frequency distributions

Number	Frequencies		Expected
	Male children	White beads	
0	5	4	6.25
1	27	34	31.25
2	64	65	62.50
3	65	70	62.50
4	30	22	31.25
5	9	5	6.25
Total	200	200	200.00
Chisquare	2.22	5.04	-

In practice, besides modern computers, natural devices like the reverse-biased diode are used to generate random numbers based on the theory of quantum mechanics which postulates the randomness of certain events at the atomic level. Note that the theory itself is verifiable by comparing the numbers so observed with sequences generated through artificial devices. However, mathematicians believe that to construct a valid sequence of random numbers (satisfying many criteria) one should not use random procedures but suitable deterministic ones! (See Hull and Dobell (1962) for an excellent discussion on this subject.) The numbers so generated are described as pseudo-random, and found to serve the desired purpose in most practical applications.

We have already seen how artificially generated random sequences of numbers enable us to discover, by comparison, similar chance mechanisms in nature and explain the occurrence of natural events such as the sequence of male and female births. There are a number of ways of exploiting randomness to make inroads on baffling questions, to solve problems that are too complex for an exact solution, to generate new information and also perhaps to help in evolving new ideas. I shall briefly describe some of them.

2.2 Monte Carlo Technique

Karl Pearson, the British mathematician and one of the early contributors to statistical theory and methods, was the first to perceive the use of random numbers for solving problems in probability and statistics that are to complex for exact solution. If you know the joint distribution of p variables, $x_1, x_2, \ldots, x_p$, how can we find the distribution of a given function $f(x_1, \ldots, x_p)$? The problem has a formal solution in the form of an incomplete multiple integral, but the computation is difficult. Pearson found random numbers useful in obtaining at least an approximate solution to such problems and encouraged L.H.C. Tippett to prepare a table of random numbers to help others in such studies. Karl Pearson said:

> The record of a month's roulette playing at Monte Carlo can afford us material for discussing the foundations of knowledge.

This method called simulation or Monte Carlo technique has now become a standard device in statistics and all sciences to solve complicated numerical problems. You generate random numbers and do simple calculations on them.

The basic principle of the simulation method is simple. Suppose that we wish to know what proportion of the area of a given square is taken up by the picture drawn inside it (see Figure 1.1). The picture has a complicated form and there is no easy way of using a planimeter to measure the area. Now, let us consider the square and take any two intersecting sides as the x and y axes. Choose a pair of random numbers (x, y) both in the range $(0, b)$ where b is greater than the length of the side of the square, and plot the point with coordinates *(x, y)* in the square. Repeat the process a number of times and suppose that at some stage, a_m is the number of points that have fallen within the picture area and m is the total number of points that have fallen within the square. There is a theorem, called the law of large numbers, established by the famous Russian probabilist A.N. Kolmogorov, which says that the ratio a_m/m tends to the true proportion of the area of the picture to that of the square, as m becomes large provided the pairs *(x, y)* chosen to locate the points are truly random. The success (or precision) of this method then depends on how faithful the random number generator is and how many we can produce subject to given resources.

Under the leadership of Karl Pearson, the method was used by some of his students to find the distribution of some very complicated sample statistics, but it did not catch up immediately except perhaps in India at the Indian Statistical Institute (ISI), where Professor P.C. Mahalanobis exploited Monte Carlo techniques, which he called random sampling experiments, to solve a variety of problems like the choice of the optimum sampling plans in survey work and optimum size and shape of plots in experimental work. The reason for delay in recognizing the potentialities of this method may be attributed to non-availability of devices to produce truly random numbers and in the requisite quantity both of which affect the precision of results. Also, in the absence of standard devices to generate random numbers, the editors of journals were reluctant to publish papers reporting simulation results. Now the situation is completely changed with the advent of reliable random number generators and easy access to them. We are able to undertake investigations of complex problems and give at least approximate solutions for practical use. The editors of journals insist that every article submitted should report simulation results even when exact solutions are available! As a matter of fact, the whole character of research in statistics, perhaps in other fields too, is gradually changing with greater emphasis on what are called "number crunching methods," of which a well known example is the

"Bootstrap method" in statistics advocated by Efron, which has become very popular. You make random numbers work.

The following diagram indicates a simple use of random numbers in estimating the area of a complicated figure.

Figure 1.1 How to find the area of a complicated figure

Monte Carlo or simulation method

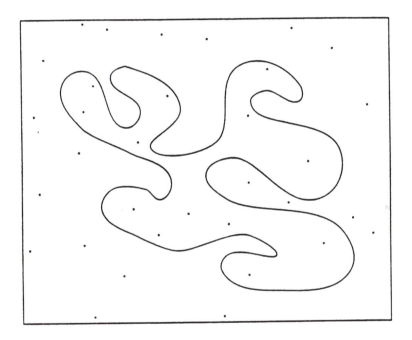

$$\frac{\text{area of the figure}}{\text{area of the square}} = \frac{\text{no. of points within the figure}}{\text{total number of random points}} = \frac{a_m}{m}$$

By the law of large numbers

$$\frac{a_m}{m} \longrightarrow \text{ true proposition as } n \rightarrow \infty$$

2.3 Sample Surveys

The next and perhaps the most important use of random numbers is in generating data in sample surveys and in experimental work. Consider a large population of individuals whose average income we wish to know. A complete enumeration or obtaining information from each individual and processing the data is not only time consuming and expensive but also in general, undesirable due to organizational difficulties in getting accurate data. On the other hand, collection of data on a small proportion of individuals (a sample of individuals) can be obtained more expeditiously and under controlled conditions to ensure accuracy of data. Then the question arises as to how the sample of individuals should be chosen to provide data from which a valid and fairly accurate estimate of the average income could be obtained. One answer is the simple lottery method using random numbers. We label all the individuals by numbers 1, 2, 3, ..., generate a certain number of random numbers in the range 1-N (where N is the total number of individuals) and select the individuals corresponding to these numbers. This is called a simple random sample of individuals.   Again, statistical theory tells us that the average of the incomes of the individuals in a random sample tends to the true value as the sample size increases. In practice, the sample size can be determined to ensure a given margin of accuracy.

2.4 Design of Experiments

Randomization is an important aspect of scientific experiments such as those designed to test whether drug A is better than drug B for treating a certain disease or to decide which is a better yielding variety of rice among a given set of varieties. The object of these experiments is to generate data which provide valid comparisons of the treatments under consideration. R.A. Fisher, the statistician who initiated the new subject of design of experiments, showed that by allocating individuals at random to the two drugs A and B in the medical experiment and assigning the varieties to the experimental plots at random in the agricultural experiment we can generate valid data for comparison of treatments.

2.5 Encryption of Messages

Random numbers in large quantities are also required in cryptology or the secret coding of messages for transmission and in maintaining the secrecy of individual bank transactions.

Top-level diplomatic and military communications, where secrecy is extremely important, are encrypted in such a way that any one illegally tapping the transmission lines can get only a random looking sequence of numbers. To achieve this, first a string of binary random digits called the key string is generated which is known only to the sender and the receiver. The sender converts his message into a string of binary digits in the usual way by converting each character into its standard eight bit computer code (the letter *a* for example is 0110 0001). He then places the message string below the key string and obtains a coded string by changing every message bit to its alternative at all places where the key bit is 1 and leaving the others unchanged. The coded string which appears to be a random binary sequence is transmitted. The received message is decoded by making the changes in the same way as in encrypting using the key string which is known to the receiver. Here is an example:

Key	0 1 0 0 0 1 1	Random digits
Message	1 0 1 1 0 0 1	Sender's message
Encrypted message	1 1 1 1 0 1 0	Transmitted message
Key	0 1 0 0 0 1 1	Same random digits
Recovered message	1 0 1 1 0 0 1	by Receiver

Banks use secret codes based on random numbers to guarantee the privacy of transactions made by automatic teller machines. For this purpose random numbers are generated as a key with a rule of converting a message into a code which is decipherable only with the knowledge of the key. Later, after the key is given to both the central computer and the teller machine, the two devices communicate freely by telephone without fear of eavesdroppers. After receiving the message from the teller machine regarding the customer's number and the amount he wants to withdraw, the central computer verifies the customer's account and instructs the teller machine to make or not make the payment.

2.6 Chance as a Tool in Model Building

The early applications of random numbers for solving statistical problems paved the way for their use in model building and prediction. Some of the areas where such models are developed are weather forecasting, demand for consumer goods, future needs of the society in terms of services like housing, schools, hospitals, transport facilities and so on. Mandelbrot (1982) provides a fascinating story of random fractals in building models of complicated curves like the irregular coast line of a country and complex shapes of natural objects.

2.7 Use in Solving Complex Problems

Some of the modern uses of random numbers which opened up a large demand for random number generators is in solving such complicated problems as the travelling salesman problem involving the determination of a minimum path connecting a number of given places to be visited starting from a given place and returning to the same place.

Another interesting example is programming the chess game. Although chess is a game with perfect information, the Artificial Intelligence (AI) programs sometimes incorporate chance moves as a way of avoiding the terrible complexity of the game.

The scope for utilization of random numbers and the concept

of chance seems to be unlimited.

2.8 Fallacies About Random Sequences

It is an interesting property of random numbers that, like the Hindu concept of God, it is patternless and yet has all the patterns in it. That is, if we go on generating strictly random numbers we are sure to encounter any given pattern sometime. Thus, if we go on tossing a coin, we should not be surprised if 1000 heads appear in successive tosses at some stage. So we have the proverbial monkey which if allowed to type continuously, can produce the entire works of Shakespeare in finite though a long period of time. (The chance of producing the drama "Hamlet" alone which has 27,000 letters and spaces is roughly unity divided by 10^{41600}. This gives some idea of how long we have to wait for the event to happen).

The patterned yet patternless nature of a sequence of random numbers has led to some misconceptions even at the level of philosophers. One is called "Gambler's Fallacy" exemplified by Polya's anecdote about a doctor who comforts his patient with the remark:

> You have a very serious disease. Of ten persons who get this disease only one survives. But do not worry. It is lucky you came to me, for I have recently had nine patients with this disease and they all died of it.

Such a view was seriously held by the German philosopher Karl Marbe (1916), who, based on a study of 200,000 birth registrations in four towns of Bavaria, concluded that the chance of a couple having a male child increases if in the past few days a comparatively large number of girls have been born.

Another, which is a counterpart of Marbe's Theory of Statistical Stabilization, is the "Theory of Accumulation" propounded by another philosopher, O. Sterzinger (1911), which formed the basis for the "Law of Series", or the tendency of the same event to occur in short runs, formulated by the biologist, Paul Kammerer (1919). A proverb says, "Troubles seldom come singly", which people take seriously and apply to all kinds of events. Narlikar (1982), in an address to the 16-th Convocation of the Indian Statistical Institute, has referred to a controversy between Fred Hoyle and Martin Ryle as arising out of such a fallacy. Professor Narlikar mentioned that his simulation or Monte Carlo experiments showed that a steady and homogeneous system can exhibit local inhomogeneities (i.e., short runs of the same event) with some frequency, and Ryle's observations of such inhomogeneities in the density of radio sources does not contradict Hoyle's steady state theory of the universe.

Let me give you another example. It is found that population sizes of a large variety of animals exhibit roughly a three year cycle, i.e., the average time that elapses between two successive peak years of population size is about 3 years. (A peak year is defined as a year in which there are more animals than in the immediately preceding and immediately succeeding years). The ubiquity of such phenomenon led some to believe that perhaps a new law of nature has been uncovered. The belief was dealt a mortal blow when it was noted that if one plots random numbers at equidistant points, the average distance between peaks approaches 3 as the series of numbers gets large. In fact, such a property is easily demonstrable by using the fact that the probability of the middle number being larger than the others in a set of 3 random numbers is 1/3. This gives an average time period of 3 years between the peaks.

2.9 Eliciting Responses to Sensitive Questions

Another interesting application of randomness is in eliciting responses to sensitive questions. If we ask a question like, "Do you smoke marijuana?", we are not likely to get a correct response. On the other hand, we can list two questions (one of which is innocuous)

S: Do you smoke marijuana?
T: Does your telephone number end with an even digit?

and ask the respondent to toss a coin and answer S correctly if head turns up and T correctly if tail turns up. The investigator does not know which question the respondent is answering and the secrecy of information is maintained. From such responses, the true proportion of individuals smoking marijuana can be estimated as shown below:

π = unknown proportion smoking marijuana, which is the parameter to be estimated.

λ = known proportion with telephone number ending with an even digit.

p = observed proportion of yes responses.

Then: $\pi + \lambda = 2p$, which provides an estimate $\hat{\pi} = 2p - \lambda$

3. From determinism to order in disorder

I shall now refer to more fundamental problems which are being resolved through the concept of randomness. These relate to building models for the universe and in framing natural laws.

For a long time it was believed that all natural phenomena have an unambiguously predetermined character, the most extreme formulation of which is to be found in Laplace's (1812) idea of a "mathematical demon", a spirit endowed with an unlimited ability for mathematical deduction, who would be able to predict all future events in the world if at a certain moment if he knew all the magnitudes characterizing its present state. Determinism, to which I have already referred, is deeply rooted in the history and prehistory of human thinking. As a concept it has two meanings. Broadly, it is an unconditional belief in the powers and omnipotence of formal logic as an instrument for cognition and description of the external world. In the narrow sense, it is the belief that all phenomena and events of the world obey causal laws. Furthermore, it implies confidence in the possibility of discovering, at least in principle, those laws from which world cognition is deduced. However, it was realized during the middle of the last century that the quest for deterministic laws of nature is strewn with both logical and practical difficulties and search for alternative models based on chance mechanisms started.

There is another aspect to Laplace's mathematical demon which is concerned with the knowledge of initial conditions of a system. It is well known that because of measurement errors, it is difficult to know the initial conditions accurately (i.e., without error). In such a case, there is a possibility that slight differences in initial conditions lead to widely different predictions for the future state of a system. A typical example is provided by Lorenz's 1961 printout of weather patterns emanating over time from nearly the same starting point. Figure 3 reproduced below from the book on Chaos by James Gleick shows how under the same law the patterns of weather, starting from initial conditions which differ in rounding off of one of the measurements .506217 to .506, grow farther and farther apart until all resemblance disappears. Such a phenomenon of sensitive dependence on initial conditions is described as the butterfly effect - the notion that a butterfly stirring the air today in Beijing can produce a storm next month in Washington.

Three major developments took place about the same time in three different fields of enquiry. They are all based on the premises that chance is inherent in nature. Adolph Qutelet (1869) used the concepts of probability in describing social and biological phenomena. Gregor Mendel (1870) formulated his laws of heredity through simple chance mechanisms like rolling a die. Boltzmann (1866) gave a special interpretation to one of the most fundamental propositions of theoretical physics, the second law of thermodynamics. The ideas propound by these stalwarts were revolutionary in nature. Although they were not immediately accepted, rapid advances took place in all these areas using statistical concepts during the present century.

The Butterfly Effect

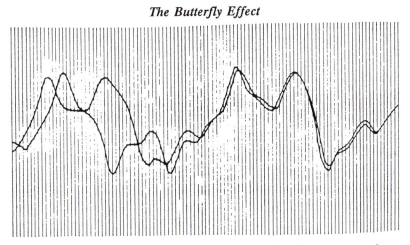

Figure 3. Graph due to Edward Lorenz showing how two weather patterns diverge from nearly the same starting point

The introduction of statistical ideas in physics started with the need to deal with errors in astronomical measurements. That repeated measurements under identical conditions do vary was known to Galileo[1] (1564-1642); he emphasized:

> Measure, measure. Measure again and again to find out the difference and the difference of the difference.

---

[1] Galileo Galilei, known by his first name was an Italian astronomer, mathematician and physicist who has been called the founder of modern experimental science. His name is associated with discoveries of the laws of pendulum, craters on the moon, sunspots, four bright satellites of Jupiter, telescope, and so on. These discoveries convinced Galileo that Nicholaus Copernicus's "Copercian Theory" that earth rotates on its axis and revolves around the sun was true. But this was contrary to Church's teaching and Galileo was forced by the Inquisition to retract his views. It is interesting to note that a few years ago, the present pope exonerated Galileo from the earlier charges made by the Church on the basis of a report submitted by a committee appointed by him.

About 200 years later, Gauss (1777-1855) studied the probability laws of errors in measurements and proposed an optimum way of combining the observations to estimate unknown magnitudes.

At a later stage, statistical ideas were used to make adjustments for uncertainties in the initial conditions and the effect of numerous uncontrollable external factors, but the basic laws of physics were assumed to be deterministic.

A fundamental change took place when the basic laws themselves were expressed in probabilistic terms especially at the microlevel of the behavior of fundamental particles. Random behaviour is considered as an "inherent and indispensable part of the normal functioning of many kinds of things, and their modes." Statistical models were constructed to explain the behavior of given systems. Examples of such descriptions are Brownian motion, the scintillations caused by radio-activity, Heisenberg's principle of uncertainty, Maxwell's velocity distributions of molecules of equal mass, etc., all of which blazed the trail for quantum mechanics of the present day. The change in our thinking is succinctly expressed by Max Born, the well known physicist,

> We have seen how classical physics struggled in vain to reconcile growing quantitative observations with preconceived ideas on causality, derived from everyday experience but raised to the level of metaphysical postulates, and how it fought a losing battle against the intrusion of chance. Today, the order has been reversed: chance has become the primary notion, mechanics an expression of its quantitative laws, and the overwhelming evidence of causality with all its attributes in the realm of ordinary experience is satisfactorily explained by the statistical laws of large numbers.

Another well known physicist A.S. Eddington goes a step further:

> In recent times some of the greatest triumphs of physical prediction have been furnished by admittedly statistical laws which do not rest on the basis

of causality. Moreover the great laws hitherto accepted as causal appear on minuter examination to be of statistical character.

The concept of statistical laws displacing deterministic laws did not find favor with many scientists including the wisest of our own century, Einstein who is quoted as having maintained even towards the end of his life:

> But I am quite convinced that some one will eventually come up with a theory, whose objects, connected by laws, are not probabilities but considered facts, as was until now taken for granted. I cannot, however, base this conviction on logical reasons, but can only produce my little finger as witness, that is, I offer no authority which would be able to command any kind of respect outside my own hand.

It is, however, surprising that Einstein accepted the chance behaviour of molecules suggested by S.N. Bose, which resulted in the Bose-Einstein theory.

Although there are uncertainties at the individual level (such as in the behaviour of individual atoms and molecules), we observe a certain amount of stability in the *average* performance of a *mass* of individuals; there appears to be "order in disorder." There is a proposition in the theory of probability, called the *Law of large numbers*, which explains such a phenomenon. It asserts that the uncertainty in the average performance of individuals in a system becomes less and less as the number of individuals becomes more and more, so that the system as a whole exhibits an almost deterministic phenomenon. The popular adage, "There is safety in numbers," has, indeed, a strong theoretical basis.

4. Randomness and creativity

We have seen how randomness is inherent in nature requiring natural laws to be expressed in probabilistic terms. We discussed how the concept of randomness is exploited in observing a small subset of a population and extracting from it information about the whole population, as in sample surveys and design of experiments. We have also seen how randomness is introduced in solving complicated problems like the travelling salesman problem and others where deterministic procedures exist but are too complex. We have also considered the use of random numbers in maintaining the secrecy of communications during transmission. Does randomness play any role in developing new ideas, or can creativity be explained through a random process?

What is creativity? It could be of different kinds. At its highest level, it is the birth of a new idea or a theory which is qualitatively different from and not conforming to or deducible from any existing paradigm, and which explains a wider set of natural phenomena than any existing theory. There is creativity of another kind at a different level, of a discovery made within the framework of an existing paradigm but of immense significance in a particular discipline. Both are, indeed, sources of new knowledge. However, there is a subtle distinction; in the first case it is created a priori though confirmed later by observed facts, and in the second case it is a logical extension of current knowledge. While we may have some idea of the mechanism behind the creative process of the second kind, it is the first kind which is beyond our comprehension. How did Ramanujan and Einstein create what they did? Perhaps we will never know the actual process, although they have some mystical explanation of creativity. However, we may characterize it in some ways.

No discovery of great importance has ever been made by logical deduction, or by strengthening the observational basis. It is then clear that a necessary condition for creativity is to let the mind wander unfettered by the rigidities of accepted knowledge or conventional rules. Perhaps, the thinking that precedes a discovery is of a fuzzy type, a successful interplay of random search for new frameworks to fit past experience and subconscious reasoning to narrow down the range of possibilities. Describing the act of creation,

Arthur Koestler says:

> At the decisive stage of discovery the codes of disciplined reasoning are suspended as they are in the dream, the reverie, the manic flights of thought, when the stream of ideation is free to drift, by its own emotional gravity, as it were, in an apparently lawless fashion.

A discovery when first announced may seem to others as without any rhyme or reason and deeply subjective. Such were indeed the reactions to the discoveries of Einstein and Ramanujan. It took a few years of experimentation and verification to accept Einstein's theory as a new paradigm, and perhaps half a century to recognize that Ramanujan's curious looking formulae have a theoretical basis of great depth and significance. Commenting on random thinking, and the role of randomness in creativity, Hofstadler says:

> It is a common notion that randomness is an indispensable ingredient of creative arts. ...Randomness is an intrinsic feature of human thought, not something which has to be artificially inseminated, whether through dice, decaying nuclei, random number tables, or what-have-you. It is an insult to human creativity to imply that it relies on arbitrary sources.

Perhaps, random thinking is an important ingredient of creativity. If this were the only ingredient, then every kind of cobweb of rash inference will be put forward and this with such great rapidity that the logical broom will fail to keep pace with them. Other elements are required such as the preparedness of the mind, ability to identify important and significant problems, quick perception of what ideas can lead to fruitful results and above all certain amount of confidence to pursue difficult problems. The last aspect is what is lacking in the bulk of scientific research today, which Einstein emphasized:

> I have little patience with scientists who take a board of wood, look for the thinnest part and drill a great number of holes where drilling is easy.

I mentioned Einstein and Ramanujan as two great creative thinkers of the present century. Perhaps it would be of interest to know a little more about their creative thought processes. To a question put to him on creative thinking, Einstein responded:

> The words or the language, as they are written or spoken, do not seem to play any role in my mechanism of thought. The physical entities which seem to serve as elements of thought are certain signs and more or less clear images which can be "voluntarily" reproduced and combined. ...this combinatory play seems to be the essential feature of productive thought - before there is any connection with logical construction in words or other kinds of signs which can be communicated to others.

Einstein was working in physics, an important branch of science. A scientific theory is valid only when its applicability to the real world is established. But when it is initiated, it is sustained by a strong faith rather than deductive or inductive reasoning. This is reflected in Einstein's dictum concerning the directness of God:

> Raffiniert ist der Herrgott, aber boshaft ist er nicht.
> (God is cunning, but He is not malicious).

Ramanujan was working in mathematics, which according to the famous mathematician Wiener is a fine art in the strict sense of the word. The validity of a mathematical theorem is in its rigorous demonstration. The proof is mathematics not so much the theorem - as mathematicians would like us to believe. To Ramanujan there are only theorems or formulae, and their validity is dictated by his intuition or his faith. He set down his formulae as works of art in patterns of supreme beauty - he said they were dictated by God in his dreams and an equation for him has no meaning unless it expresses a thought of God. God, beauty and truth are perceived to be the same. We would not have had Ramanujan if he did not believe in this.

Ramanujan recorded a large number of theorems in a Note Book during his last year of life from his death bed. This Note Book which was discovered a few years ago has a number of conjectures one of which is as follows:

Some conjectures (formulae) in the Lost Note Book of Ramanujan.

Professor G. Andrews, who brought the Lost Note Book to light informs me that the equality in the first three lines of the formula (known as mock theta conjecture) is recently proved by D.R. Hickerson of the Penn State University.

References

Boltzmann, L. (1910): *Vorlusungen Uber Gastheorie*, 2 vols, Leipzig.

Efron, B. and Tibshirani, R.J. (1993): *An Introduction to the Bootstrap*, Chapman & Hall.

Gleick, James (1987): *Chaos*, Viking, New York, p.17

Hull, T.E. and Dobell, A.R. (1962): Random number generators, *SIAM Rev.*4, 230.

Kammerer, P. (1919): Das Gasetz der Serie, eine Lehre van den Wiederholungen im Labensund im Welteshehen, Stuttgart and Berlin.

Laplace, P.S. (1914): Essai philodophique de probabilitis, reprinted in his *Théorie analytique des probabiliés* (3rd ed. 1820).

Mahalanobis, P.C. (1954): The foundations of statistics, *Dialectica* **8**, 95-111.

Mandelbrot, B.B. (1982): *The Fractal Geometry of Nature,* W.H. Freeman and Company, San Francisco.

Marbe, K. (1916): *Die Gleichförmigkeit in der Weit, Utersuehungen zur Philosophie and positiven Wissenschaft,* Munich.

Mendel, G. (1870): *Experiments on Plant Hybridization* (English Translation) Harvard University Press, Cambridge, 1946.

Narlikar, J.V. (1982): Statistical techniques in astronomy, *Sankhyā* **42**, 125-134.

Quetelet, A. (1869): *Physique aociale ou essai sur le dévelopment des facultés de l'homme,* Brussels, Paris, St. Petersburg.

Sterzinger, O. (1911): *Zur Logic and Naturphilosophie der Wahrscheintichkeitslehre,* Leipzig.

Tippett, L.H.C. (1927): *Random Sampling Numbers.* Tracts for computers, No.15 Ed. E.S. Pearson, Camb. Univ. Press.

Appendix: Discussion

A.1 *Chance and Chaos*

During the discussion after a talk I gave based on the material of this Chapter, a question was raised about chaos, a term used to describe "random like" phenomena, and its relation to the study of chance and uncertainty. My response was as follows.

The word chance is used to describe random phenomena like drawing of numbers in a lottery. A sequence of numbers so produced do exhibit some order in the long run which can be explained by the calculus of probability. On the other hand, it is observed that numbers produced by a deterministic process may exhibit randomlike behaviour locally while having a global regularity. During the last 20 years scientists have started studying the latter type of phenomena under the name chaos. A new approach is suggested to model complex shapes and forms such as cloud formation, turbulence, coast line of a country, and even to explain variations in stock market prices by the use of simple mathematical equations. This way of thinking is somewhat different from invoking a chance mechanism to describe the outcomes of a system. *Chance deals with order in disorder while chaos deals with disorder in order.* Both may be relevant in modeling observed phenomena.

The study of chaos came into prominence with the discovery of Edward Lorenz of what is called the "Butterfly Effect", or sensitive dependence of a system on initial conditions. He observed that in long range weather forecasting, small errors in initial measurements used as inputs in the prediction formula may give rise to extremely large errors in predicted values. Benoit Mandelbrot invented Fractal Geometry to describe a family of shapes which exhibit the same type of variation at different scales. His Fractal Geometry could explain shapes which are "jagged, tangled, splintered, twisted and fractured" as we find in nature, such as the formation of snowflakes and coast line of a country. Mitchell J. Feigenbaum developed the concept of strange attractors based on iterated functions,

$$x, \; f(x), \; f(f(x)), \; \ldots$$

which provide an accurate model for several physical phenomena such as fluid turbulence.

The chaos that scientists are talking about is mathematical in nature and its study is made possible and attractive by the use of computers. It is a pastime which paid well and opened up new ways of modeling observed phenomena in nature through deterministic models.

An interesting example due to the famous mathematician Mark Kac (see his autobiography *Enigmas of Chance*, pp.74-76) shows how the graph of a deterministic function could mimic the tracing of a random mechanism. To test Smoluchowski's theory of Brownian motion of a little mirror suspended on a quartz fiber in a vessel containing air, Kappler conducted an ingenious experiment in 1931 to obtain photographic tracings of the motion of the mirror. One such a tracing of 30 seconds' duration is reproduced in the figure below.

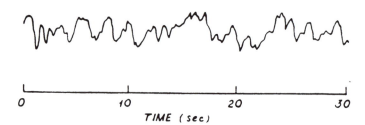

Kac remarks that looking at the graph, "it is difficult to escape the feeling that one is in the presence of chance incarnate and the tracing could only have been produced by a random mechanism." Kappler's experiment might be interpreted as confirming Smoluchowski's theory that the mirror is hit at random by the molecules of air giving the graph of the displacement of the mirror the character of a stationary Gaussian process.

Kac shows that the same kind of tracing indistinguishable from Kappler's graph by any statistical analysis, can be produced by plotting the function given below

$$\alpha \; \frac{\cos \lambda_1 t + \cos \lambda_2 t + \ldots + \cos \lambda_n t}{\sqrt{n}}$$

for sufficiently large n, choosing a sequence of numbers $\lambda_1, \ldots, \lambda_n$ and a scale factor α. Kac asks: So what is chance?

A.2 Creativity

Dr. J.K. Ghosh, Director of the Indian Statistical Institute has sent me the following comments.

There is always something mysterious and awe-inspiring about creativity, and there is more of this in Ramanujan's work than any one else can think of in the twentieth century. Reflecting on the nature of this mysterious element in the act of creation, that is, in the birth of new ideas or new discoveries, Professor Rao speculates whether randomness is not an important part of creativity. In fact he puts forward a new tentative paradigm for understanding creativity. Let me quote from him. "It is then clear that a necessary condition for creativity is to let the mind wander unfettered by the rigidities of accepted knowledge or conventional rules. Perhaps the thinking that precedes a discovery is a fuzzy type, a successful interplay of random search for new frameworks to fit past experience and subconscious reasoning to narrow down the range of possibilities." Perhaps even the random search is at a subconscious level. That much creative work gets done at a subconscious level has been authenticated many times - a brilliant account was complied by Hadamard [Hadamard, J. (1954): Essay on the psychology of invention, in the book *Mathematical Field*, Princeton, Dover.] But the association with randomness and uncertainty, the concepts that we quantify through probability statements, is a brilliant additional hypothesis. In the form of a vague reference to chance it occurs in Hadamard, but does not receive much attention. It is probably the central thesis to which Professor Rao leads us through a dizzying glimpse of Ramanujan's almost magical powers and a masterly overview of randomness and uncertainty. The following remarks are confined to this thesis.

It seems to me that there is always an element of creativity, including its magical quality, when one makes an inductive leap or even when one is involved in a non-trivial learning process. Two consequences would seem to follow from this. First, at least part of the mystery relating to creativity is related to a lack of proper philosophical foundations for induction, in spite of many attempts, specially by the Viennese school. Such attempts have been frivolously described as attempts to pull a very big cat out of a very small bag. Secondly the mystery of creativity is also related to a lack of a satisfactory model for learning in artificial intelligence. A third fact, which is relevant in this context is worth pointing out. As far as I know, the only models of learning, at least of adaptive learning, are stochastic. It would then seem to follow that Professor Rao's hypothesis is a brilliant but logical culmination of such modeling. If one were to try to make a computer do creative work, i.e., simulate creativity, this seems to be the only way of doing it at present. I wonder whether music generated on computers is of this sort.

How satisfactory, illuminating or acceptable can such models be? In this connection I would like to refer to Hilbert's paradigm for mathematics. Today the essence of mathematics seems to be best understood by familiarizing oneself with Hilbert's program for finitistic formalism as well as Godel's impossibility theorem. (There are optimistic exceptions, for example, Nelsen (*Sankhyā*, A, 1985).) Creativity, like induction, is too complex to generate even an impossibility theorem. It makes sense to talk of impossibility only when it refers to a precisely defined algorithm. However, one could probably have examples which are in some sense counterintuitive with respect to a given model. Then the model along with the "counter examples" may help one grasp better the nature of what is being modeled. I feel such counter examples exist with respect to Professor Rao's hypothesis but, in justification, can only fall back on a statement of Einstein which Professor Rao himself quotes: "I cannot however base this conviction on logical reasons, but can only produce my little finger as witness." Dr. Ghosh concludes his comment by saying: I don't know if my views are a sort of Popperism about creativity. I don't know Popper's views about science well enough for that.

I thank Dr. Ghosh for raising some fundamental issues on the much debated concept of creativity. I restrict my reply to creativity in science which is perhaps different from that in music, literature and arts (see Chandrasekhar 1975: The Nora and Edwary Rayerson lecture on Shakespeare, Newton and Beethoven on Patterns of Creativity). In science, the bulk of research work done is at the level of mopping up operations, plugging a hole or caulking a leak. There is a small percentage of research which is identifiable as creative and which itself may be at two levels of sophistication: that made within the framework of an existing paradigm and that, at a higher level, involving a paradigm shift. The mechanism of creative processes of both kinds may not be completely known, but few aspects of it are generally recognized: subconscious thinking when the mind is not constrained by logical deductive processes, serendipity, transferring experience gained in one area to a seemingly different area, and even aesthetic feeling for beauty and patterns. The following is a sample of quotations about creativity.

> *pour inventor il faut penser à côté.*
> *(to invent you must think aside)* — Souriau

> *One sometimes finds what one is not looking for.* — A. Fleming

> *I do not seek, I find.* — Picasso

> *My work always tried to unite the true with the beautiful; but when I had to choose one or the other, I usually chose the beautiful.* — H. Weyl

> *I had my results for a long time, but I do not yet know how to arrive at them.* — Johann Gauss

> *Hypotheses non fingo (I frame no hypotheses)* — Isaac Newton

> *I have said that science is impossible without faith. ...Inductive logic, the logic of Bacon, is rather something on which we can act than something which we can prove, and to act on it is a supreme assertion of faith...Science is a way of life which can only flourish when men are free to have faith.* — Norbert Wiener

There is a certain element of mysticism in the initiation of creative science reflected in the above quotations. Some philosophers discussed the issue of creativity without throwing much light on it.

With reference to Popper's views referred to by Dr. Ghosh I

may say the following. Popper's statements that scientific hypotheses are just conjectures can only be interpreted to mean that hypotheses formulation from observational facts has no explicit algorithm. Popper's assertion that a hypothesis cannot be accepted but can only be falsified may have deep philosophical significance but is not valid in its strict sense, as scientific laws are, in fact, applied in practice successfully. Popper does not attach any importance as to how hypotheses are formulated. It may be, because there is no logical answer to such question even if it is raised.

I believe that scientific laws which have an impact on science can be built on existing knowledge and/or induction alone. It demands a creative spark of "imagining things that do not exist and asking why not" (in the words of George Bernard Shaw). I suggested random thinking as an ingredient of creativity. At the stage of intensive activity of the human brain trying to resolve a problem, "when all the brain cells are stretched to the utmost," some random moves away from conventional thinking may be necessary to discover a plausible solution. This does not mean that the search for a solution is made by random trial and error from a possible set of finite alternatives. In a creative process, the alternatives are not known in advance. They may not be finite. I am referring to the final steps in a discovery process where optimum choices are made sequentially, based on the knowledge gained by previous choices and possibilities are narrowed down till what is believed to be a reasonable choice emerges. It is a process (perhaps a stochastic one) of gradually dispelling darkness and not one of deciding which out of a possible set of windows could be opened to throw most light. However, there are some scientists who believe that computers can be exploited in the creation of new knowledge.

To what extent can creativity be mechanized? In the context of scientific discoveries, some experimental studies have been made to demonstrate that a scientific discovery, however revolutionary it may be, comes within the normal problem-solving process and does not involve mythical elements associated with it such as "creative spark", "flash of genius" and "sudden insight". As such it is believed that creativity results from information processing and hence programmable.

In a recent book *Scientific Discovery* (Computational Exploration of Creative Processes, MIT Press, Cambridge), the authors, Pat Langley, Herbert A. Simon, Gary L. Bradshaw and Jan M. Zytkow, discuss the taxonomy of discovery and the possibility of writing computer programs, for information processing aimed at "problem finding," "identification of relevant data" and "selective search guided by heuristics," the major ingredients of creativity. They have given examples to show that several major discoveries made in the past could have been accomplished, perhaps more effectively through computer programs using only the information and knowledge available at the times of these discoveries. The authors hope that the theory they have on problem solving will provide programs to search for solutions even involving paradigm shifts initiating new lines of research. The authors conclude by saying:

> We would like to imagine that the great discoverers, the scientists whose behaviour we are trying to understand, would be pleased with this interpretation of their activity as normal (albeit high-quality) human thinking. ...Science is concerned with the way the world is, not with how we would like it to be. So we must continue to try new experiments, to be guided by new evidence, in a heuristic search that is never finished but it always fascinating.

A similar sentiment about the nature of science is expressed by Einstein:

> Pure logical thinking cannot yield us knowledge of the empirical world. All knowledge of reality starts from experience and ends with it. Propositions arrived at by purely logical means are completely empty of reality.

But the role of the mind in a creative process is emphasized by Roger Penrose in his book *The Emperor's New Mind*:

> The very fact that the mind leads us to truths that are not computable
> convinces me that a computer can never duplicate the mind.

A.3 *Chance and Necessity*

During the discussion, questions were raised about cause and effect and chance occurrences, which may be summarized as follows: "You have emphasized the uncertainty of natural events. If events happen at random, how can we understand, explore and explain nature?"

I am glad this question is raised. Life would be unbearable if events occur at random in a completely unpredictable way and uninteresting, in the other extreme, if everything is deterministic and completely predictable. Each phenomenon is a curious mixture of both, which makes "life complicated but not uninteresting" (as J. Neyman used to say).

There are logical and practical difficulties in explaining observed phenomena and predicting future events through the principle of cause and effect.

Logical, since we can end up in a complex cause-effect chain. If A_2 causes A_1, we may ask what causes A_2. Say A_3; then what causes A_3 and so on. We may have an endless chain and at some stage the quest for a cause may become difficult or even logically impossible forcing us to model events at that stage through a chance mechanism.

Practical, since, except in very trivial cases, there are infinitely many (or finitely large number of) factors causing an event. For instance, if you want to know whether the toss of a coin results in a head or a tail you must know several things. First, the magnitudes of numerous factors such as initial velocity (x_1), measurements of the coin (x_2), nervous state of the individual tossing the coin (x_3), ..., which determine the event (y), head or tail, and then the relationship

$$y = f(x_1, x_2, x_3, \ldots)$$

must be known. Uncertainty arises if f is not known exactly, if the values of all the factors $x_1, x_2, \ldots$ cannot be ascertained and if there are measurement errors. We may have information only on some of the factors, say $x_1, \ldots, x_n$, forcing us to model the outcome y as

$$y = f_a(x_1, \ldots, x_n) + e$$

where f_a is an approximation to f and *e* is the unknown error arising out of our choice of f_a, lack of knowledge on the rest of the factors and measurement errors. Modeling for uncertainty in the choice of f_a and the error ϵ through a chance mechanism becomes a necessity.

What is chance and how to model it? How do we combine the effects of known causes with the possible effects of unknown causes in explaining observed phenomena or predicting future events? What are meant by "explaining a phenomenon" and "prediction of an event" when there is uncertainty? Indeed, there are logical difficulties in answering such questions. If we are modeling uncertainty, the question of modeling uncertainty in modeling uncertainty would naturally arise. We may set aside these philosophical issues and interpret an explanation of a phenomenon as a working hypotheses (not absolutely true) from which deductions can be drawn within permissible margins of error.

The first attempt in this direction is the development of theory of errors, where uncertainties in measurements have to be taken into account in interpreting results (estimating unknown quantities and verifying hypotheses). The second stage is the characterization of observed phenomena in terms of laws of chance governing a physical system. This is probably the greatest advance in human thinking and understanding nature, and a striking example is the work of Gregor Mendel who introduced, for the first time, 120 years ago the *indeterministic paradigm* in the history of science. He laid the foundations of genetics, the hereditary mechanism, by observing data subject to chance fluctuations. Mendel's ideas led to the modern theory of evolution, which is a "mixture of chance and necessity - chance at the level of variation and necessity in the working of selection." Then came the breakthrough in explaining physical phenomena through random behaviour of fundamental particles. The concept of chance has actually helped in unraveling the mystery behind what appeared to be happening without a cause.

We have also gone ahead and learnt to deal with chance in any given situation whether it arises in our daily life, scientific research, industrial production or complex decision making. We have developed methods to extract signals from messages distorted by chance events (noise) and to reduce chance effects through feedback and control (cybernetics, servomechanism). We have devised methods for peaceful coexistence with chance, methods that enable us to work effectively despite the presence of chance effects (use of error correcting codes, repeating experiments for consistency, introducing redundancy to enable easy recognition). Most amazing of all, we have learnt to utilize chance to solve problems which are otherwise difficult to solve (Monte Carlo, random search) and to make improvements (selection in breeding programs). An element of chance is sometimes deliberately incorporated in the design of machines by engineers to enhance their performance. Most paradoxical of all, we artificially introduce chance elements in the collection of data (as in sample surveys and design of experiments) to provide valid and unbiased information.

The full impact of the acceptance of the Dice-Playing God running the universe has yet to come. As Rustum Roy says (in his book *Experimenting with Truth*, p.188):

> The planning of society at community and national levels must be shaped differently to keep in resonance with the bell shaped curve of the "normal" distribution under which we all live.

He goes on to say that one profound political consequence may be the abolition of election processes of campaigning by candidates (self selling) and voting by people and introduction of selection by a random process (lottery method) from a set of qualified persons.

I would like to recall what L. Rastrigin, Director of the only Random Research Laboratory in the world, located in Russia, mentioned in his popular book, *The Chancy, Chancy World*:

> The study of the remarkable world of chance is only just beginning. Science has as yet barely skimmed the surface of this world of strange happenings and limitless potential. But the excavation of the priceless treasures of chance has begun, and there is no telling what riches it may yet uncover. One thing, however, is certain: we shall have to get used to thinking of chance, not as an irritating obstacle, not as an "inessential adjunct to phenomena" (as it is in the philosophical dictionary), but as a source of unlimited possibilities of which even the boldest imagination can have no prescience.

If we were to speak of any rational principle in nature, then that principle can only be chance: for, it is chance, acting in collaboration with selection, that constitutes nature's "reason". Evolution and improvement are impossible without chance.

A.4 *Ambiguity*

Besides chance and randomness which we discussed earlier, there is another obstacle in interpreting observed data. This is *ambiguity* in identifying objects (persons, places or things) as belonging to distinct categories. Am I a statistician, a mathematician or an administrator? I may give different answers in different situations. Occasionally, I may say that I am one-third of each. Of course, it is essential to define categories with as much precision as possible to avoid confusion in communicating our ideas and in investigation work. But ambiguity in introducing concepts and making definitions cannot be avoided. "That there is no God-given way to establish categories, much less place people in them, is a fundamental difficulty" (Kruskal, 1978, private communication). I believe the need to study "fuzzy sets" in mathematics arose out of ambiguity in identification of objects.

However, it is interesting to note that Edward Levi in his classic 1949 book on legal reasoning writes at length about the important role of ambiguity in the court and the legislature. Kruskal (1978) gives the following quotations from Levi's book to highlight the above theme.

> The categories used in the legal process must be left ambiguous in order to permit the infusion of new ideas. (p.4)

> It is only folklore which holds that a statute if clearly written can be completely unambiguous and applied as intended to a specific case. Fortunately or otherwise, ambiguity is inevitable in both statute and constitution as well as with case law. (p.6)

> [Ambiguity in legislature] is not the result of inadequate draftsmanship, as often urged. ...Even in a non-controversial atmosphere just exactly what has been decided will not be clear. ...[It is necessary] that there be ambiguity before there can be any agreement about how known cases will be handled. (pp.30-31)
> This is the only kind of system which will work when people do not agree completely. ...The words change to receive the content which the community gives to them. (p.104)

Thus for Dr.Levi, ambiguity is not a dragon, but beneficent and necessary for coherence of society.

It appears that two essential elements which make life interesting are chance and ambiguity - unpredictability of natural events and the lack of unique interpretation of the terms we use in communication. In the past, both are considered as obstacles about which nothing can be done. We are now learning not only to accept them as ineluctable, but, perhaps, consider them as essential for the progress of our society!

A.5 *Are decimal digits in π random?*

In an article published in the *International Statistical Review, Vol 64, 329-344, 1996,* Y. Dodge traces the 4000-year old history of π and raises the question whether the decimal digits of π form a random sequence. Technically speaking, a random sequence of symbols is a sequence which cannot be recorded by means of an algorithm in a form shorter than the sequence itself. In this strict sense the sequence of decimal digits in π don't form a random sequence. It is interesting to note that computers are being used to find the decimal places of π using a version of Ramanujan's mysterious formula

$$\frac{1}{\pi} = 2\sqrt{2} \sum_{n=0}^{\infty} \frac{\left[\frac{1}{4}\right]_n \left[\frac{1}{2}\right]_n \left[\frac{3}{4}\right]_n}{(1)_n (1)_n n!} (1103 + 26390n) \left[\frac{1}{99}\right]^{4n+2}$$

However, the decimal digits of π may be described as pseudo random numbers as they satisfy all known statistical tests of randomness. As such they can be used in all simulation studies to derive valid results which are as good as those obtained by using randomly generated numbers by the lottery method.

The first 1000 decimal digits[2] of π are given in Table 1.4. The frequencies of the numbers 0,1,...,9 in the 1000 decimal digits are as follows:

Digits	0	1	2	3	4	5	6	7	8	9
Frequency	93	116	103	102	93	97	94	95	101	106
Expected	100	100	100	100	100	100	100	100	100	100

---

[2] It is reported that a Chinese boy Zhang Zuo aged 12 could recall from memory the first 4000 digits of π in 25 minutes and 30 seconds.

Table 1.4 The first 1000 decimal digits of π

1415926535	8979323846	2643383279	5028841971	6939937510
5820974944	5923078164	0628620899	8628034825	3421170679
8214808651	3282306647	0938446095	5058223172	5359408128
4811174502	8410270193	8521105559	6446229489	5493038196
4428810975	6659334461	2847564823	3786783165	2712019091
4564856692	3460348610	4543266482	1339360726	0249141273
7245870066	0631558817	4881520920	9628292540	9171536436
7892590360	0113305305	4882046652	1384146951	9415116094
3305727036	5759591953	0921861173	8193261179	3105118548
0744623799	6274956735	1885752724	8912279381	8301194912
9833673362	4406566430	8602139494	6395224737	1907021798
6094370277	0539217176	2931767523	8467481846	7669405132
0005681271	4526356082	7785771342	7577896091	7363717872
1468440901	2249534301	4654958537	1050792279	6892589235
4201995611	2129021960	8640344181	5981362977	4771309960
5187072113	4999999837	2978049951	0597317328	1609631859
5024459455	3469083026	4252230825	3344685035	2619311881
7101000313	7838752886	5875332083	8142061717	7669147303
5982534904	2875546873	1159562863	8823537875	9375195778
1857780532	1712268066	1300192787	6611195909	2164201989

The value of the chisquare statistic for testing departure of the observed frequencies from the expected is 4.20, which is small for 9 degrees of freedom, indicating close agreement with equal frequencies. Another test is to consider the frequencies of odd digits in sets of five decimal digits, which are as follows:

No. of odd digits	0	1	2	3	4	5
Frequency	7	31	54	61	41	6
Expected	6.25	31.25	62.50	62.50	31.25	6.25

The chisquare for testing agreement with expected values is 4.336 which is small for 5 degrees of freedom. The sequence of decimal digits of π seems to share the same property as the sequence of male and female births or of white and black beads illustrated in Tables 1.1 and 1.2 in Section 2.1.

Chapter 2

Taming of Uncertainty-Evolution of Statistics

The quiet statisticians have changed our world - not by discovering new facts or technical developments but by changing the ways we reason, experiment, and form our opinions about it.

- Hacking

1. Early history: Statistics as data

Statistics has a long antiquity but a short history. Its origin could be traced back to the beginning of mankind, but only in recent times it has emerged as a subject of great practical importance. At present, it is a lively subject, widely used despite controversies about its foundations and methodology. There have been fashions in statistics advocated by different schools of statisticians. The advent of computers is having considerable impact on the development of statistical methodology under the broader title of data analysis. It is not clear what the future of statistics will be. I shall give a brief survey of the origin of statistics, discuss the current developments and speculate on its future.

1.1 *What is statistics?*

Is statistics a separate discipline like physics, chemistry, biology, or mathematics? A physicist studies natural phenomena like heat, light, electricity and laws of motion. A chemist determines compositions of substances and interactions between chemicals, and a biologist studies plant and animal life. A mathematician indulges in his own game of deducing propositions from given postulates. Each of these subjects has *problems of its own* and *methods of its own* for

41

solving them, which gives it the status of a separate discipline. Is statistics a separate discipline in this sense? Are there purely statistical problems which statistics purports to solve? If not, is it some kind of art, logic or technology applied to solve problems in other disciplines?

A few decades ago, statistics was neither a frequently used nor a well understood word. Often it was viewed with skepticism. There were no professionals called statisticians except a few employed in government departments to collect and tabulate data needed for administrative purposes. There were no systematic courses at the universities leading to academic degrees in statistics. Now the situation is completely changed. Statistical expertise is in great demand in all fields of human endeavor. A large number of statisticians are employed in government, industry and research organizations. The universities started teaching statistics as a separate discipline. All these phenomenal developments raise a number of questions:

* What is the origin of statistics?
* Is statistics a science, technology or art?
* What is the future of statistics?

1.2   *Early Records*

The earliest record of statistics is, perhaps, notches on trees cut by primitive man, even before the art of counting was perfected to keep an account of his cattle and other possessions. The need for collection of data and recording information must have arisen when human beings gave up independent nomadic existence and started living in organized communities. They had to pool their resources, utilize them properly and plan for future needs. Then came the establishment of kingdoms ruled by kings. There is evidence that rulers of ancient kingdoms all over the world had accountants who collected detailed data about the people and the resources of the state.

One of the early Chinese emperors Liu Pan considered statistics so important that he made his Prime Minister in charge of statistics, a tradition which continued for a long time in China. It was in their interests to know how many able-bodied men might be mobilized in times of emergency, how many would be needed for essentials of civil life; how numerous or how wealthy were certain minorities who might resent some contemplated changes in the laws of property or of marriage; what was the taxable capacity of Province, their own and of their neighbors.

There is evidence that as early as 2000 B.C., during the time of Hsia Dynasty, censuses were taken in China. In Chow Dynasty (1111 B.C. - 211 B.C.), an official position entitled "Shih-Su" (bookkeeper) was established to take charge of statistical work. In the book *Kuan Tzu*, Chapter 24 was entitled *Inquiry* in which sixty-five questions were carried to deal with every aspect in governing a state. For example, how many households owned land and houses? How much food stock did a family have? How many widowers, widows, orphans, disabled and sick people were there?

The fourth book of the Old Testament contains references to early censuses conducted about 1500 B.C., and instructions to Moses to conduct a census of the fighting men of Israel.

The word census itself was derived from the Latin word *censere*, which means to tax. The Roman census was established by the sixth king of Rome, Servius Tullius (578-534 B.C.). Under this system, Roman officials called *censors* made a register at 5-year intervals of the people and their property for taxation purposes and for determining the number of able-bodied fighting men. In 5 B.C., Caesar Augustus extended the census to include the entire Roman Empire. The last regular Roman census was conducted in 74 A.D. There is no record of census taking anywhere in the Western World for several centuries after the fall of the Roman Empire. Regular periodic censuses as we know them today started only in the seventeenth century.

It is interesting to know that in India a very elaborate system of what we now call administrative records or official statistics was evolved before 300 B.C. In the text *Arthaśāstra* of Kautilya, published sometime in the period 321-300 B.C. (see subsection 2.3), there is a detailed description of how data should be collected and recorded. Gopa, the village accountant, was required to maintain all kinds of records about the people, land utilization, agricultural produce, etc. An example of his duties mentioned in *Arthaśāstra* was as follows:

> Also having numbered the houses as taxpaying or nontax paying, he shall not only register the total number of inhabitants of all the four castes in each village, but also keep an account of the exact number of cultivators, cowherds, merchants, artisans, laborers, slaves, and biped and quadruped animals, fixing at the same time the amount of gold, free labor, toll and fines that can be collected from it (each house).

In recent times, under the Mohammedan rulers of India, we find official statistics occupying a very important place. The most well known compilation of this period is *Ain-i-Akbari*, the great administrative and statistical survey of India under Emperor Akbar which was completed by the minister Abul Fazl in 1596-97 A.D. It contains a wealth of information regarding a great empire, of which a random selection is as follows:

> Average yield of 31 crops for 3 different classes of land; annual records of rates based on the yield and price of 50 crops in 7 provinces extending over 19 years (1560-61 to 1578-79 A.D.); daily wages of men employed in the army and the navy, laborers of all kinds, workers in stables etc.; average prices of 44 kinds of grains and cereals, 38 vegetables, 21 meats and games, 8 milk products, oils, and sugars, 16 spices, 34 pickles, 92 fruits, 34 perfumes, 24 brocades, 39 silks, 30 cotton clothes, 26 woolen stuffs, 77 weapons and accessories, 12 falcons, elephants, horses, camels, bulls and cows, deer, precious stones, 30 building materials, weights of 72 kinds of wood etc.

It is not clear why and how such masses of data were compiled, what administrative machinery was used, what precautions were taken to ensure completeness and accuracy, and for what purpose they were used.

1.3   *Statistics and Statistical Societies*

The term *statistics* has its roots in the Latin word *status* which means the state, and it was coined by the German scholar Gottfried Achenwall about the middle of the eighteenth century to mean

collection, processing and use of data by the state.

In his book, *Elements of Universal Erudition* published in 1770, J. von Bielfeld refers to statistics as

the science that teaches us what is the political arrangements of all the modern states of the known world.

The *Encyclopedia Britannica* (third edition, 1979) mentions statistics as

a word lately introduced to express a view or a summary of any kingdom, country or parish.

About this time, the word "publicistics" was also used as an alternative to statistics but its usage was soon given up. C.A.V. Malchus amplifies the scope of statistics in his book, *Statistic und Staatskunde* published in 1826, as

the most complete and the best grounded knowledge of the condition and the development of a given state and of the life within it.

In Britain, Sir John Sinclair used the word statistics in a series of volumes issued during 1791-1799, entitled "The Statistical Account of Scotland: an enquiry into the state of the country for the purpose of ascertaining the quantum of happiness enjoyed by its inhabitants and the means of its future improvement." It was said that the British expressed their surprise at Sir John's using German words "statistics" and "statistical" instead of using equivalent words in the English language.

Thus, to the political arithmeticians of the eighteenth century, statistics was the science of statecraft - its function was to be the eyes and ears of the government.

However, the raw data are usually voluminous and confusing, They have to be suitably summarized for easy interpretation and possible use in making policy decisions. The first attempts in this direction were made by a prosperous London tradesman, John Graunt (1620-1674), in analyzing the Bills of Mortality (lists of the dead with the cause of death). He produced a pamphlet where he had "reduced several great confusing volumes (of Bills of Mortality) into a few perspicuous tables, and abridged such observations as naturally flowed from them, in a few succinct paragraphs, without any long series of multiloquious deductions". He drew useful conclusions on issues such as the relative death rates from various diseases and on the growth of populations in the countryside and in the city of London. He also constructed life tables which laid the foundations of the subject of Demography. John Graunt was, thus, a pioneer in demonstrating the use of statistics in describing the current state of affairs and in guiding the future course of events.

The next steps in the application of statistics to human affairs were taken by the Belgian mathematician Adolphe Quetlet (1796-1874). Under the influence of Laplace, he studied probability and developed an interest in statistics and its application to human affairs. He collected all sorts of social data and described the frequency distribution in terms of the normal law, which he called "the law of accidental causes." In 1844, Quetlet astonished skeptics by using the normal law of distribution of heights of men to discover the extent of draft evasion in France. By comparing the distribution of heights of those who answered the call for the draft with the actual distribution of heights in the general population he computed that about 2000 men had evaded conscription by pretending to be less than the minimum height. He showed how to forecast future crimes of different kinds by studying the past trends. To promote the study of statistics and encourage its use in making policy decisions, he urged Charles Babbage (1792-1871) to found the statistical society of London (1834). Then he made the Crystal Palace Expositions in London in 1851 a forum for international cooperation, which only three years later produced the First International Statistical Congress (1854) at Brussels. As the first president, he preached the need for uniform procedures and terminology in compiling statistical data. Quetlet tried to establish statistics as a tool in improving society. The modern concepts of economics and demography such as GNP (Gross National Product), rates of growth and development, and population growth are a legacy of Quetlet and his disciples.

Statistics appears to have been recognized as a science when it was included as a section in the British Association for the Advancement of Science, and the Royal Statistical Society was founded in 1834. By then, statistics was considered as

> facts relating to men, which are capable of being expressed in numbers, which sufficiently multiplied, to indicate general laws.

With the rapid industrialization of Europe in the first half of the nineteenth century, public interest began to be aroused in questions relating to the conditions of the people. In this period, particularly in the years 1830-1850, statistical societies were founded in some countries, and statistical offices were set up in many countries for the purpose of "procuring, arranging and publishing facts calculated to illustrate the conditions and prosperity of the society." [France established the Central Statistical Bureau in 1800, the first one in the world.] In this context it was natural to inquire how each country was developing in relation to the others with a view to determine factors responsible for growth. For such useful analytical studies, it was necessary to have data collected from different countries on a comparable basis. This was sought to be achieved by arranging international congresses periodically to agree upon concepts and definitions and uniform methods of collection of data, "thus enhancing the value of all future observations by making them more comparable as well as more expeditiously collected." The first congress was held in Brussels in 1853, which was attended by 153 delegates representing 26 countries. A series of other congresses followed, each emphasizing the need for agreement among different Governments and Nations to undertake "common inquiries, in a common spirit, by a common method for a common end."

It was clear that if statistics were to be useful and developed as a tool for research, international cooperation was necessary. For exchanging experience and setting up common standards, a number of international congresses of statistics (about 10) were held during the period 1853-1876 at the invitation of different countries in Europe. As these congresses were found to be useful, a proposal was made at the golden jubilee celebration of the Statistical Society of London held in 1885 to establish an International Statistical Society to follow up on the resolutions passed at each congress and to lay down plans for future congresses. After some discussions it was resolved to establish a permanent organization to be called the International Statistical Institute. Thus the ISI was born on June 24, 1885. The rules and regulations of the Institute prescribed among other things the holdings of biennial sessions, the nature of membership, publication of journals, etc. The main emphasis was placed in achieving "uniformity in methods of compiling and abstracting statistical returns and in inviting the attention of the governments to the use of statistics in solving problems." A permanent office of the Institute was later established at the Hague in 1913 to look after the publications of the Institute.

The ISI has considerably expanded its activities over the last 100 years. Separate associations for mathematical statistics and probability, statistical computing, sample surveys, official statistics and statistical education were formed within the ISI administrative set up of ISI.

2.   Taming of uncertainty

As I have already said, statistics in the original etymological sense comprises the activities of collection and compilation of data and their possible use in public policy making.

During the nineteenth century, statistics began to acquire a new meaning as interpretation of data or methods of extracting information from data for making decisions. How can we make forecasts of socio-economic characteristics of a population based on current trends? What is the effect of certain legislation adopted by the government? How do we make policy decisions to increase the welfare of the society? Can we develop a system for insuring against failure of crops, death and catastrophic events?

There are others questions awaiting satisfactory answers. Will it rain tomorrow? How long will the current warm spell last or, at a more scientific level, do observed data support a given theory? At a personal level, there arise questions of the type: What prospects do I have in the career I have chosen? How do I invest my money to maximize the return?

The main obstacle in answering questions of these types is *uncertainty* - lack of one to one correspondence between cause and effect. How does one act under uncertainty? This has baffled mankind for a long time and it is only in the beginning of the present century that we have learnt to tame uncertainty and develop the science of *wise decision making*. Why did it take such a long time for the human mind to come up with solutions to perplexing problems confronting us every moment of our lives? To answer this question, let us examine the logical processes or types of reasoning we usually employ in solving problems and creating new knowledge, and the changes that have taken place in our thinking over the last twenty-five centuries.

2.1 *Three Logical Types of Reasoning*

2.1.1 Deduction

Deductive reasoning was introduced by the Greek philosophers more than two thousand years ago and perfected over the last several centuries through the study of mathematics. We have given premises or axioms, say A_1, A_2, A_3, ..., each of which is accepted to be true by itself. We can choose any subset of the axioms, say A_1, A_2, to prove a proposition P_1. The truth of P_1 solely depends on the truth of the axioms A_1, A_2; the fact that the other axioms are not explicitly used in the argument has no relevance. Similarly using A_2, A_3, A_4 we may derive a proposition P_2 and so on.

By deductive reasoning no new knowledge is created beyond the premises, since all the derived propositions are implicit in the axioms. There is no claim that either the axioms or the derived propositions have any relation to reality as characterized by the following quotations.

> Mathematics is a subject in which we do not know what we are talking about, nor care whether what we say is true.   - Bertrand Russel

> The mathematician may be compared to a designer of garments who is utterly oblivious of the creatures whom his garments may fit.
>
> - Tobias Dantzig

It is interesting to note that deductive logic which is the basis of mathematics considered to be the "highest truth" is not without logical flaws. As observed earlier, in deductive logic it is permissible to prove a proposition choosing any subset of the axioms and the fact that other axioms are not used has no relevance.

Then the following question arises. Is it possible that one subset of axioms say A_1, A_2 imply the proposition P and another subset A_2, A_3, A_4 imply the proposition not P, leading to a

Deductive Reasoning

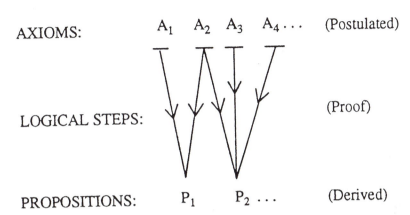

AXIOMS: A_1 A_2 A_3 A_4 . . .   (Postulated)

LOGICAL STEPS: (Proof)

PROPOSITIONS: P_1 P_2 . . .   (Derived)

Can P_1 and P_2 be contradictory?

contradiction? Can it happen that postulate A_1, A_2 imply that the sum of three angles of a triangle is 180 while postulates A_3, A_4, A_5 imply some other number? Attempts to prove that no such contradiction arises with the axioms of mathematics has resulted in some surprises. Gödel, the famous mathematical logician put forward an ingenious proof, by an elaborate argument, to the effect that you could not basing your reasoning on a given set of axioms disprove the possibility that the system could lead to a contradiction.

It was also established that if a system of axioms allows the deduction of a particular proposition P as well as not P, then the system of axioms enables us to derive any contradiction we like. I would like to recall an anecdote mentioned by Sir Ronald Fisher in his lecture on "Nature of Probability" published in *The Centennial Review*, Vol.11, 1958. G.H. Hardy, the famous British mathematician remarked on this remarkable fact at the dinner table one day in Trinity College, Cambridge. A Fellow sitting across the table took him up.

Fellow: Hardy, if I said that 2+2=5, could you prove any other proposition you like?

Hardy: Yes, I think so.

Fellow: Then prove that McTaggart is the Pope.

Hardy: If 2+2=5, then 5=4. Subtracting 3 from each side 5-3=4-3, i.e., 2=1.

McTaggart and the Pope are two, but two is one. Therefore McTaggart is the Pope.

Mathematics is a game played with strict rules, but there is no knowing whether some day it will be found to be a bundle of inconsistencies.

2.1.2 Induction

The story is different with inductive reasoning. Here we are confronted with the reverse problem of deciding on the premises given some of its consequences. It is the reasoning by which decisions are taken in the real world based on incomplete or shoddy information. Some examples where induction is necessary are as follows:

Making decisions under uncertainty in a unique situation

* Did the accused in a given case commit the murder?
* Is the mother's allegation that a particular person fathered her child true?

Prediction

* It has been continuously raining in State College from Monday to Friday. Will it continue to rain in the weekend?
* What will be the drop in the Dow Jones index tomorrow?
* What is the demand for automobiles next year?

Testing of hypothesis

* Is Tylenol better than Bufferin in relieving headache?
* Does eating oat bran cereal reduce cholesterol?

These are some of the situations in the real world where decisions have to be taken under uncertainty. We have observed data which could have resulted from any one of a set of possible hypotheses or causes, i.e., the correspondence between data and hypothesis is not one to one. Inductive reasoning is the logical process by which we match a hypothesis to given data and thus generalize from the particular. This way, we are creating new knowledge, but it is uncertain knowledge because of lack of one to one correspondence between data and hypothesis. This lack of precision in our inference from given data, unlike in deductive inference from given axioms, stood in the way of codifying inductive reasoning. To the human mind accustomed to deductive logic, the concept of developing a theory or introducing rules of reasoning which need not always give correct results must have appeared unacceptable. So, inductive reasoning remained more as an art with a degree of success depending on an individual's skill, experience and intuition.

Inductive Reasoning

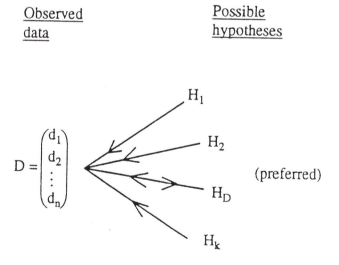

* Can we lay down rules for preferring one or a subset of hypotheses based on given data?
* What is the uncertainty in the choice of a particular hypothesis H_D made by following a specified rule of procedure?

2.1.3 The logical equation of risk management

The breakthrough came only in the beginning of the present century. It was realized that although the knowledge created by any rule of generalizing from the particular is uncertain, it becomes certain knowledge, although of a different kind, once we can quantify the amount of uncertainty in it. The new paradigm is the logical equation:

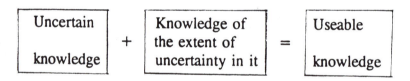

$$\text{Uncertain knowledge} + \text{Knowledge of the extent of uncertainty in it} = \text{Useable knowledge}$$

This is not philosophy. This is a new way of thinking. This is a basic equation which has led to an efficient way of risk management and liberated humanity from the oracles and sooth sayers. It puts the future at the service of the present replacing helplessness by judicious decision making:

* If we have to take a decision under uncertainty, mistakes cannot be avoided.

* If mistakes cannot be avoided, we better know how often we make mistakes (knowledge of amount of uncertainty) by following a particular rule of decision making (creation of new but uncertain knowledge).

* Such a knowledge could be put to use in finding a rule of decision making which does not betray us too often or which minimizes the frequency of wrong decisions, or which minimizes the loss due to wrong decisions.

The problem as formulated of optimum decision making can be solved by deductive reasoning. Thus inductive inference could be brought within the realm of deductive logic.

For example, let us consider the form in which weather forecasts are made nowadays. Not long ago, weather predictions used to be in the form of categorical statements like: It will rain tomorrow. It will not rain tomorrow. Obviously, such forecasts could be wrong a large number of times. Now, they make forecasts like: There is 30% chance of rain tomorrow, which may appear as a noncommittal statement. How is the number 30% arrived at. A friend of mine, a mathematician, says that at the TV station they have 10 meteorologists and each one is asked whether it rains tomorrow or not. If three of them say yes, then it is announced that there is 30% chance of rain tomorrow.

Of course, this is not how the figure 30% is arrived at. It has a deeper meaning. It represents the frequency of occasions on which it rained in the past on the next day when the atmospheric conditions on the previous day were as observed today. It tells us the *amount* of uncertainty about rain tomorrow and is based on complex modelling of weather patterns and computations carried out on a vast mass of observed data. In this sense, the statement made about tomorrow's weather in terms of chances of rain is a precise one, as precise as a mathematical theorem, and conveys all the necessary information for an individual to plan his activities for the next day. Different individuals would use this information in different ways to their benefit. A categorical statement such as it will rain tomorrow without a measure of uncertainty in the statement is of no practical value. In some sense it is illogical.

Table 2.1 Weather Forecast (quantification of uncertainty)

Data	Possibilities	Chances
Today's atmospheric conditions	It will rain tomorrow	30%
	It will not rain tomorrow	70%

There is a noticeable difference between deduction and induction. In deductive inference, it is permissible to choose a subset of premises to prove a proposition. In inductive inference, different subsets of data may lead to different and often contradictory conclusions and *it is therefore imperative that all data must be used.* Editing or rejection of data, if necessary, must be dictated by the process of inference and not by a choice of the data analyst.

The statement that we can prove anything by statistics only means that we can always select a portion of available data to support any preconceived idea. This is what politicians and sometimes scientists do to sell their pet ideas or what the business men manipulate to sell their products.

There is another aspect of inductive inference which is worth noting. It is important that we use only the given data and not any unverified assumptions or preconceived notions as inputs. Let us look at the sad plight of a prince who believed that only maids are employed in a royal palace:

> The prince, travelling through his domains, notices a man in the cheering crowd who bore a striking resemblance to himself. He beckoned him over and asked, "Was your *mother* ever employed in my palace?
> "No Sire," the man replied, "But my *father* was."

2.1.4 Abduction

Sometimes new theories are proposed without any data base

purely by intuition or flash of imagination, which is called *abduction* in logical terminology. They are verified later by conducting experiments. Famous examples of this are double helix nature of DNA, theory of relativity, electromagnetic theory of light and so on.

The distinction between induction and abduction is somewhat subtle. In induction, we are guided by experimental data and its analysis to provide an insight. But the ultimate step in the creation of new knowledge does depend to some extent on previous experience and a flight of imagination. This led to the belief that all induction is abduction.

To summarize, advancement of knowledge depends on these logical processes:

Induction: Creation of new knowledge based on observational data.

Abduction: Creation of new knowledge by intuition without data base.

Deduction: Verification of proposed theories.

2.2   *How to quantify uncertainty?*

The main concept which has led to codification of inductive inference is quantification of uncertainty, as in the case of weather prediction illustrated in Table 2.1. The 30% chance of rain tomorrow was based on previous observations. There is, however, no definite way of doing this and the subject is full of controversies. It has even created different schools of statisticians advocating different ways of quantifying uncertainty.

The first serious attempt to quantify uncertainty was made by Reverend Thomas Bayes (?-1761) who was said to be 59 years old at the time of his death. [The year of his birth is unknown.] He introduced the concept of a *prior* distribution on the set of possible hypotheses, indicating perhaps our degree of beliefs for different hypotheses before data are observed. We denote this by p(h) and consider it as given. This together with a knowledge of the probability distribution of data (d) given a hypothesis (h), denoted by p(d|h), enables us to obtain the total (marginal) probability distribution of observed data denoted by p(d). We are now in a position to compute the conditional probability distribution of hypotheses given data, called Bayes theorem,

$$p(h|d) \ = \ \frac{p(h)p(d|h)}{p(d)}$$

which is the *posterior distribution* or the distribution of uncertainties about the alternative hypotheses in the light of the observed data. From a prior knowledge of the alternative hypotheses and observed data, we have obtained new knowledge about possible hypotheses.

Bayes theorem is an ingenious attempt in using the theory of probability as an instrument in inductive reasoning. However, some statisticians feel somewhat uneasy about the introduction of a prior distribution p(h) in a problem, unless the choice of such a distribution is made objectively, for instance, based on past observational evidence, and not on one's beliefs or mathematical convenience in computing the posterior distribution. Indeed, attempts were made by the founders of modern statistics, K. Pearson (March 27, 1857 - April 27, 1936), R.A. Fisher (February 17, 1890 - July 29, 1962), J. Neyman (April 16, 1894 - August 5, 1981), E.S. Pearson (August 11, 1895 - June 12, 1980) and A. Wald (October 31, 1902 - December 13, 1950), to develop theories of inference without the use of a prior distribution. These methods are not without logical difficulties. However, the lack of a fully logical methodology has not prevented the use of statistics in day to day decision making or for unraveling the mysteries of nature. The situation is similar to what we have in medicine; you do not hesitate to treat a patient with an available drug even though it is not the ideal remedy or it has undesirable side effects or, in rare cases, its efficiency is not fully established by field studies. But search for new drugs must continue. The methodology of statistics developed in the first half of this century for the estimation of unknown parameters, testing of hypotheses and decision making has opened up the flood gates for applications in many areas of human endeavor and the need for forging new tools for dealing with uncertainty is increasing rapidly. Statistics has excelled any technology or scientific invention of the present century in its ubiquity and in opening the gates to new knowledge.

With quantification of uncertainty, we are able to raise new questions which cannot be answered by the classical or Aristotelian logic based on two alternatives "yes" or "no", and provide solutions for practical applications. We are able to manage individual and institutional activities in an optimum way by controlling, reducing and, what is more important, making allowance for uncertainty. There is wisdom in what Descartes (1596-1650) said more than three hundred years ago,

> It is truth very certain that when it is not in our power to determine what is true we ought to follow what is most probable.

Thus the new discipline of extracting information from data and drawing inferences was born and the scope of the term statistics was extended from just data to interpretation of data.

To sum up, chance is no longer something to worry about or an expression of ignorance. On the contrary, it is the most logical way to present our knowledge. We are able to come to terms with uncertainty, to recognize its existence, to measure it and to show that advancement of knowledge and suitable action in face of uncertainty are possible and rational. As Sir David Cox put it:

> Recognition of uncertainty does not imply nihilism; nor need it force us into what Americans sometimes call one-handedness.

Chance may be the antithesis of all law. But the way out is to discover the laws of chance. We look for the alternatives and provide the probabilities of their happening as measures of their uncertainties.

Knowing the consequences of each event and the probability of its happening, decision making under uncertainty can be reduced to an exercise in deductive logic. It is no longer a hit and miss affair.

3. Future of Statistics

Statistics is more a way of thinking or reasoning than a bunch of prescriptions for beating data to elicit answers.

Is statistics, as studied and practiced today, a science, technology or art? Perhaps it is a combination of all these.

It is a science in the sense that it has an identity of its own with a large repertoire of techniques derived from some basic principles. These techniques cannot be used in a routine way; the user must acquire the necessary expertise to choose the right technique in a given situation and make modifications, if necessary. Statistics plays a major role in establishing empirical laws in soft sciences. Further, there are philosophical issues connected with the foundations of statistics - the way uncertainty can be quantified and expressed - which can be discussed independently of any subject matter. Thus in a broader sense statistics is a separate discipline, perhaps a discipline of all disciplines.

It is a technology in the sense that statistical methodology can be built into any operating system to maintain a desired level and stability of performance, as in quality control programs in industrial production. Statistical methods can also be used to control, reduce and make allowance for uncertainty and thereby maximize the efficiency of individual and institutional efforts.

Statistics is also an art, because its methodology which depends on inductive reasoning is not fully codified or free from controversies. Different statisticians may arrive at different conclusions working with the same data set. There is usually more information in given data than what can be extracted by available statistical tools. Making figures tell their own story depends on the skill and experience of a statistician, which makes statistics an art, as in the example of the Red Fort Story (Section 2.14, Chapter 5).

What is the future of statistics? Statistics is now evolving as a metascience. Its object is the logic and the methodology of other sciences - the logic of decision making and the logic of experimenting in them. The future of statistics lies in the proper communication of statistical ideas to research workers in other branches of learning; it will depend on the way the principal problems are formulated in other fields of knowledge.

On the logical side, the methodology of statistics is likely to be broadened for using expert evidence in addition to information supplied by data in assessment of uncertainty.

Having said that statistics is science, technology as well as an art - the newly discovered logic for dealing with uncertainty and making wise decisions - I must point out a possible danger to its future development. I have said earlier that statistical predictions could be wrong, but there is much to be gained by relying on statistically predicted values rather than depending on hunches or superstitious beliefs. Can the customer for whom you are making the prediction sue you if you are wrong? There have been some recent court cases. I quote from an editorial of *The Pittsburgh Press*, dated Saturday, May 24, 1986 under the title, Forecasters Breathe Easier:

> A federal appeals court has wisely corrected a gross miscalculation of government liability in a case involving weather forecasting.
>
> Last August, a U.S. District judge awarded $1.25 million to the families of three lobster-men who were drowned during a storm that had not been predicted. The judge said the government was liable because it had failed to repair promptly a wind sensor on a buoy used to help forecast weather conditions off Cape Cod.
>
> The award was overturned the other day by the appeals court on grounds that weather forecasting is a "*discretionary function* of government and *not a reliable one at that*".
>
> "Weather predictions fail on frequent occasions" the appeals court said. "If in only a small proportion of cases, parties suffering in consequence succeeded in producing an expert who could persuade a judge

that the government should have done better," the burden on the government "would be both *unlimited* and *intolerable.*"

The case isn't over yet, since it probably will be appealed to the Supreme Court. But government meteorologists practicing their *inexact science* are breathing a bit easier.

Such instances will be rare, but none-the-less may discourage statistical consultants from venturing into new or more challenging areas and restrict the expansion of statistics.

Chapter 3

Principles and Strategies of Data Analysis: Cross Examination of Data

1. Historical developments in data analysis

Data! data! he cried impatiently,
I can't make bricks without clay.
                    Conan Doyle - *The Copper Beeches*

Styles in statistical analysis change over time while the object of "extracting all the information from data" or "summarization and exposure" remains the same. Statistics has not yet aged into a stable discipline with complete agreement on foundations. Certain methods become popular at one time and are replaced in course of time by others which look more fashionable. In spite of controversies, the statistical methodology and fields of applications are expanding. The computers together with the availability of graphic facilities have had a great impact on data analysis. It may be of interest to briefly review some historical developments in data analysis.

It has been customary to consider descriptive and theoretical statistics as two branches of statistics with distinct methodologies. In the former, the object is to summarize a given data set in terms of certain "descriptive statistics" such as measures of location and dispersion, higher order moments and indices, and also to exhibit salient features of the data through graphs such as histograms, bar diagrams, box plots and two dimensional charts. No reference is made to the stochastic mechanism (or probability distribution) which gave rise to the observed data. The descriptive statistics thus computed are used to compare different data sets. Even some rules are prescribed for the choice among alternative statistics, such as the mean, median and mode, depending on the nature of the data set and the questions to be answered. Such statistical analysis is referred to as descriptive data analysis (DDA). In theoretical statistics, the object is again summarization of data, but with reference to a specified family (or model) of underlying probability distributions. The summary or descriptive statistics in such a case heavily depend on the specified stochastic model, and their distributions are used to specify margins of uncertainty in inference about the unknown parameters. Such methodology is referred to as inferential data analysis (IDA).

Karl Pearson (K.P.) was the first to try to bridge the gap between DDA and IDA. He used the insight provided by the descriptive analysis based on moments and histograms to draw inference on the underlying family of distributions. For this purpose he invented the first and perhaps the most important test criterion, the chi-squared statistic, to test the hypothesis that a given data arose from a specified stochastic model (family of probability distributions) or consistent with a given hypothesis, which "ushered in a new sort of decision making" [See Hacking (1984), where K.P.'s chi-squared is eulogized as one of the top 20 discoveries[1] since 1900 considering all branches of science and technology. Even R.A. Fisher (R.A.F.) who had personal differences with K.P. expressed his appreciation of K.P.'s chi-squared test in personal conversation with the author.] K.P. also created a variety of probability distributions distinguishable by four moments. A beautiful piece of research work done by K.P. through the use of histograms and chi-squared test is the discovery that the distribution of the size of trypanosomes found in certain animals is a mixture of two normal distributions (see Pearson (1914-15)).

The need to develop general methods of estimation arose in applying the chi-squared test to examine a composite hypothesis that

---

[1] The top 20 discoveries considered are, in no particular order: Plastics, the IQ test, Einstein's theory of relativity, blood types, pesticides, television, plant breeding, networks, antibiotics, the Taung skull, atomic fission, the big-bang theory, birth control pills, drugs for mental illness, the vacuum tube, the computer, the transistor, statistics (what is true and what is due to chance), DNA, and the laser.

the underlying distribution belongs to a specified parametric family of distributions. K.P. proposed the estimation of parameters by moments, and using the chi-squared test based on the fitted distribution. Certain refinements were made by R.A.F. both in terms of obtaining a better fit to given data through the estimation of unknown parameters by the method of maximum likelihood and also in the exact use of the chi-squared test using the concept of degrees of freedom when the unknown parameters are estimated.

During the twenties and thirties, R.A.F. created an extraordinarily rich array of statistical ideas. In a fundamental paper in 1922 he laid the foundations of "theoretical statistics," of analyzing data through specified stochastic models. He developed exact small sample tests for a variety of hypotheses under normality assumption and advocated their use with the help of tables of certain critical values, usually 5% and 1% quantiles of the test criterion. During this period, under the influence of R.A.F., great emphasis was laid on tests of significance and numerous contributions were made by Hotelling, Bose, Roy and Wilks among others to exact sampling theory. Although R.A.F. mentioned specification, the problem first considered by K.P., as an important aspect of statistics in his 1922 paper, he did not pursue the problem further. Perhaps in the context of small data sets arising in biological research which R.A.F. was examining, there was not much scope for investigating the problem of specification or subjecting observed data to detailed descriptive analysis to look for special features or to empirically determine suitable transformations of data to conform to an assumed stochastic model. R.A.F. used his own experience and external information of how data are ascertained in deciding on specification. [See the classical paper by R.A.F. (1934) on the effect of methods of ascertainment on the estimation of frequencies.] At this stage of statistical developments inspired by R.A.F.'s approach, attempts were made by others to look for what are called nonparametric test criteria whose distributions are independent of the underlying stochastic model for the data (Pitman, 1937) and to investigate robustness of test criteria proposed by R.A.F. for departures from normality of the underlying distribution.

The twenties and thirties also saw systematic developments in data collection through design of experiments introduced by R.A.F., which enabled data to be analyzed in a specified manner through analysis of variance and interpreted in a meaningful way: design dictated the analysis and analysis revealed the design.

While much of the research in statistics in the early stages was motivated by problems arising in biology, parallel developments were taking place in a small scale on the use of statistics in industrial production. Shewhart (1931) introduced simple graphical procedures through control charts for detecting changes in a production process, which is probably the first methodological contribution to detection of outliers or change points.

Much of the methodology proposed by R.A.F. was based on intuition, and no systematic theory of statistical inference was available except for some basic ideas in the theory of estimation. R.A.F. introduced the concepts of consistency, efficiency and sufficiency and the method of maximum likelihood in estimation. J. Neyman and E.S. Pearson in 1928 (see their collected papers) provided some kind of axiomatic setup for deriving appropriate statistical methods, especially in testing of hypotheses, which was further pursued and perfected by Wald (1950) as a theory for decision making. R.A.F. maintained that his methodology was more appropriate in scientific inference while conceding that the ideas of Neyman and Wald might be more relevant in technological applications, although the latter claimed universal validity for their theories. Wald also introduced sequential methods for application in sampling inspection, which R.A.F. thought had applications in biology also. [In an address delivered at the ISI, R.A.F. mentioned, Shewhart's control charts, Wald's sequential sampling and sample surveys as three important developments in statistical methodology.]

The forties saw the development of sample surveys which involved collection of vast amounts of data by investigators by eliciting information from randomly chosen individuals on a set of questions. In such a situation, problems such as ensuring accuracy (free from bias, recording and response errors) and comparability (between investigators and methods of enquiry) of data assumed paramount importance. Mahalanobis (1931, 1944) was perhaps the first to recognize that such errors in survey work were inevitable and could be more serious than sampling errors, and steps should be taken to control and detect these errors in designing a survey and to develop suitable scrutiny programs for detecting gross errors (outliers) and inconsistent values in collected data.

We have briefly discussed what are commonly believed to be two branches of statistics, viz., descriptive and inferential statistics, and the need felt by practicing statisticians to clean up the data of possible defects which may vitiate inferences drawn from statistical analysis. What was perhaps needed is an integrated approach, providing methods for a proper understanding of given data, its defects and special features, and for selection of a suitable stochastic model or a class of models for analysis of data to answer specific questions and to raise new questions for further investigation. A great step in this direction was made by Tukey (1962, 1977) and Mosteller and Tukey (1968) in developing what is known as exploratory data analysis (EDA). The basic philosophy of EDA is to understand the special features of data and to use robust procedures to accommodate for a wide class of possible stochastic models for the data. Instead of asking the Fisherian question as to what summary statistics are appropriate for a specified stochastic model, Tukey proposed asking for what class of stochastic models, a given summary statistic is appropriate. Reference may also be made to what Chatfield (1985) describes as initial data analysis, which appears to be an extended descriptive data analysis and inference based on common sense and experience with minimal use of traditional statistical methodology.

The various steps in statistical data analysis are exhibited in Chart 1, which is based on my own experience in analyzing large data sets and which seems to combine K.P.'s descriptive, Fisher's

Chart 1
Various steps in Statistical Data Analysis

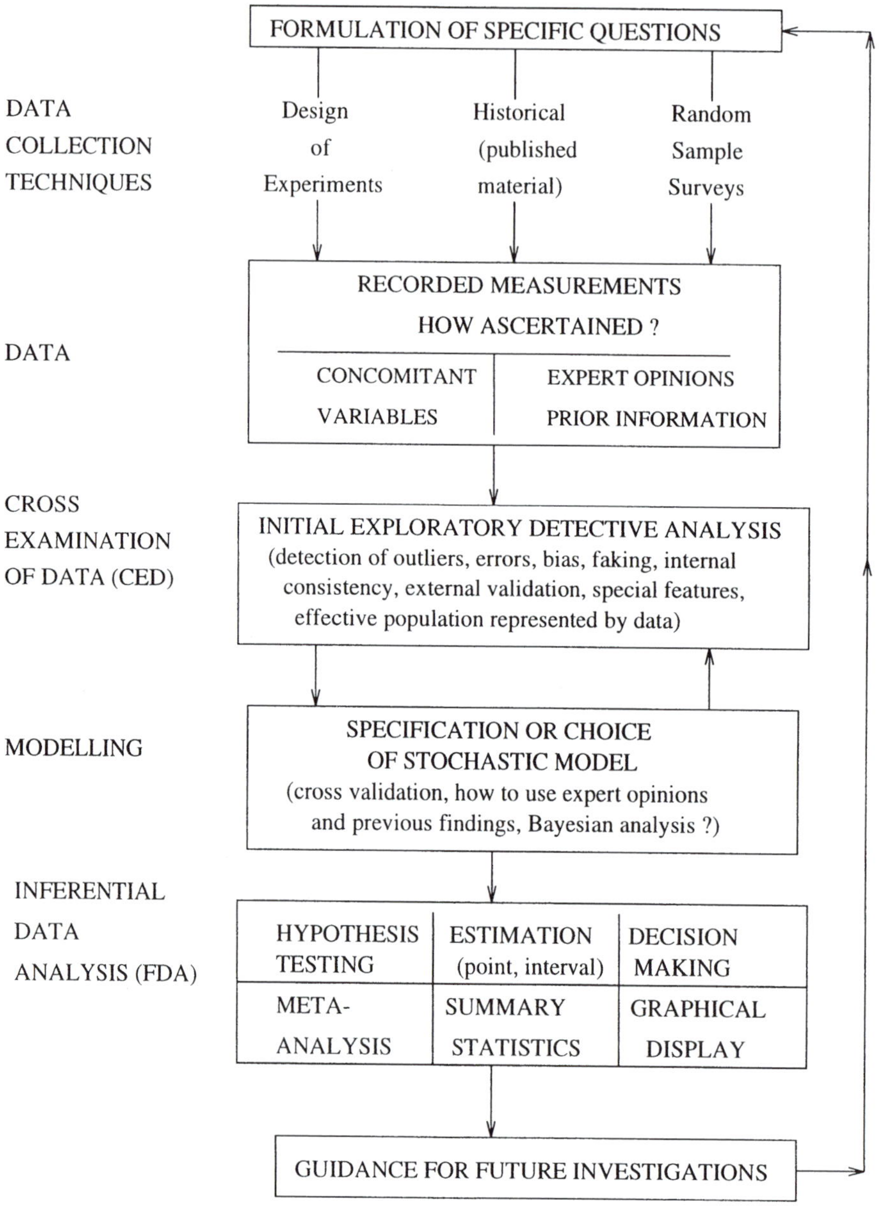

	FORMULATION OF SPECIFIC QUESTIONS	

DATA
COLLECTION
TECHNIQUES

Design of Experiments	Historical (published material)	Random Sample Surveys

DATA

RECORDED MEASUREMENTS HOW ASCERTAINED ?	
CONCOMITANT VARIABLES	EXPERT OPINIONS PRIOR INFORMATION

CROSS
EXAMINATION
OF DATA (CED)

INITIAL EXPLORATORY DETECTIVE ANALYSIS
(detection of outliers, errors, bias, faking, internal
consistency, external validation, special features,
effective population represented by data)

MODELLING

SPECIFICATION OR CHOICE
OF STOCHASTIC MODEL
(cross validation, how to use expert opinions
and previous findings, Bayesian analysis ?)

INFERENTIAL
DATA
ANALYSIS (FDA)

HYPOTHESIS TESTING	ESTIMATION (point, interval)	DECISION MAKING
META- ANALYSIS	SUMMARY STATISTICS	GRAPHICAL DISPLAY

GUIDANCE FOR FUTURE INVESTIGATIONS

inferential and Tukey's exploratory data analyses, and Mahalanobis' concern for non-sampling errors.

In Chart 1, data is used to represent the entire set of recorded measurements (or observations) and how they are obtained, by an experiment, sample survey or from historical records, and the operational procedures involved in recording the observations, and any prior information (including expert opinions) on the nature of data or the stochastic model underlying the data.

Cross-examination of data (CED) represents whatever exploratory or initial study is done to understand the nature of the data, to detect measurement errors, recording errors and outliers, to test validity of prior information and to examine whether data are genuine or faked. The initial study is also intended to test the validity of a specified model or select a more appropriate stochastic model or a class of stochastic models for further analysis of data.

Inferential data analysis (IDA) stands for the entire body of statistical methods for estimation, prediction, testing of hypotheses and decision making based on chosen stochastic model for observed data. The aim of data analysis should be to extract all available information from data and not merely confined to answering specific questions. Data often contain valuable information to indicate new lines of research and to make improvements in designing future experiments or sample surveys for data collection. I would like to enunciate the main principle of data analysis in the form of a fundamental equation:

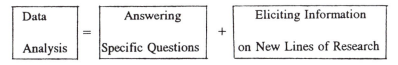

$$
\begin{array}{|c|}
\hline
\text{Data} \\
\\
\text{Analysis} \\
\hline
\end{array}
=
\begin{array}{|c|}
\hline
\text{Answering} \\
\\
\text{Specific Questions} \\
\hline
\end{array}
+
\begin{array}{|c|}
\hline
\text{Eliciting Information} \\
\\
\text{on New Lines of Research} \\
\hline
\end{array}
$$

The sequence of data analysis indicated in Chart 1 as CED and IDA should not be regarded as distinct categories with different methodologies. It only shows what we should do to begin with when presented with data and in what form the final results should be expressed and used in practical applications. Some results of IDA may suggest further CED, which in turn may indicate changes in IDA.

An important aspect of data analysis is that no extraneous assumptions, not supported by present data or past experience, should be used as inputs. A question has been raised as to the role of expert opinions in data analysis. My answer is:

> Use expert opinions in such a way that we stand to gain if they are correct and do not lose if they are wrong.

Thus expert opinions would be useful in the planning stage of a survey or designing an experiment.

2.   Cross-examination of data

Figures won't lie, but liars can figure.
General Charles H. Grosvenor

Statisticians are often required to work on data collected by others. The first task of a statistician, as Fisher put it, is cross examination of data (CED), the art of making figures speak, to obtain all the information necessary for a meaningful analysis of data and interpretation of results. A possible check list for CED under broad categories with specific items under each category is as follows.

* How are the data ascertained and recorded?

* Are the data free from measurement and recording errors? Are the concepts and definitions associated with measurements well defined? Are there differences between observers?

* Are the data genuine, i.e., ascertained as stated, or

faked or edited or adjusted in any way? Are any observations discarded at the discretion of the observer? Are there any outliers in the data which might have undue influence in statistical inference?

* What is the effective population for which the observed data provides information? Is there any non-response (partial or complete) from selected units of a population under survey? Are the data obtained from a homogeneous or a mixture of populations? Are all relevant factors for identification and classification of sampled units recorded?

* Is there any prior information on the problem under investigation or on the nature of observed data?

Answers to some of these questions may be had by talking to the investigator who collected the data; but for the rest, answers may have to be elicited through appropriate analysis of data, i.e., by addressing the questions to data or cross examining the data. This is not a routine matter although graphical representation of data through histograms, two dimensional scatter plots, and probability plots of suitably transformed measurements, and computation of certain descriptive statistics would be of great help. However, much depends on the nature of the data and the skill of a statistician to elicit information from the data (to make figures speak). I shall consider some examples.

2.1 *Editing of data*

Let us look at the following table which appears on page 74 in the book "Epidemiology, Man and Disease" by J.P. Fox, C.E. Hall and L.R. Elveback.

The authors conclude that "although the attack rates are high

Table 3.1 The number of people attacked with measles and of those that died, by age groups, during the epidemic of measles in Faroe Islands in 1846

Age (Years)	Population	Number Attacked	Attack Rate (percent)	Number of Deaths	Case Fatality (percent)
<1	198	154	77.8	44	28.6
1-9	1440	1117	77.7	3	0.3
10-19	1525	1183	77.6	2	0.2
20-29	1470	1140	77.6	4	0.3
30-39	842	653	77.6	10	1.5
40-59	1519	1178	77.6	46	3.9
60-79	752	583	77.5	46	7.9
80+	118	92	78.0	15	16.3
Total	7864	6100	77.6	170	2.8

Source: Peter L. Panum. Observations Made During the Epidemic of Measles on the Faroe Islands in the Year 1846. New York: Delta Omega Society, 1940, p.82.

in all age groups, the fatality varied significantly, being higher under one year and then rising steadily for those over age thirty." Is this conclusion valid?

What is striking in the table is the rather uniform attach rates of measles for all age groups (indicated by blocking) with very little or no variation from the overall attack rate of 77.6. Could this occur by chance even if the true attack rate is common to all the age groups? There is a strong suspicion that the number attacked in each age group was not observed but reconstructed from the known population size in each age group by multiplying it by the common overall attack rate of $6100/7864 = .776$ and rounding off to the nearest integer. Thus the figures 154 for age less than 1, and 92 for over 80, could have been obtained as follows:

$$198 \times .776 = 153.648 \sim 154; \quad 118 \times .776 = 91.568 \sim 92 \qquad (2.1.1)$$

Now, if we use these reconstructed round numbers to calculate the attack rates we get the values

$$\frac{154}{198} = .777\ldots \sim .778; \frac{92}{118} = .7796 \sim .780 \qquad (2.1.2)$$

as reported by the authors and also explains why the reported attack rates differ slightly in the third decimal place. A reference to the original report in German by the well known German epidemiologist who was sent to Faroe Islands to combat the epidemic of measles, Panum revealed that the number attacked was not originally classified by age groups but the number attacked in each group was reconstructed in the manner explained in the equation (2.1.1) by the editor of the English translation assuming a uniform attack rate. The attack rates reported in the blocked column of the above table are not found in the table on page 87 of the English translation, which are probably computed by the authors Fox, Hall and Elveback of the book, *Epidemiology, Man and Disease,* in the manner explained in (2.1.2). In view of this, the age specific fatality rates computed from the reconstructed values of the number attacked in each group and the consequent interpretation may not be valid. A statistician is often required to do detective type of work! (The second entry in the blocked column should be 77.6!)

2.2   *Measurement and recording errors, outliers*

In any large scale investigation, measurements and recording errors are inevitable. It is difficult to detect them unless they appear as highly discordant values not in line with the others. Care should be taken to see in designing an investigation that such errors are minimized. A built-in scrutiny program while measurements are being made in the field might alert the investigator when a value looks suspicious and allow him to repeat the measurement and/or investigate whether or not the individual being measured belongs to the population under study.

The author had the opportunity to scrutinize vast amounts of data collected in anthropometric surveys. In one case, the entire data collected at great cost had to be rejected (see Mukherji, Rao and Trevor (1955) and Majumdar and Rao (1958)). When the number of recording and measurement errors in multivariate response data is not large, they could be detected by drawing histograms of individual measurements and ratios, plotting two dimensional charts for pairs of measurements, and computing the first four moments and measures of skewness and kurtosis, γ_1 and γ_2. The last two measures are specially sensitive to outliers. Table 3.2 gives the values of γ_1 and γ_2 computed from the original data and after removing extreme values for a number of characteristics for different populations sampled. The sample size for each group was of the order of 50. The asterisks indicate significance at the 5% level. It is seen that the recomputed values of γ_1 and γ_2 after omitting one extreme value in each case, are in conformity with others.

Table 3.2 Test statistics γ_1 for skewness and γ_2
for kurtosis for some anthropometric measurements of
five male tribal populations
(Source: Ph.D. Thesis of Urmila Pingle)

| Charac- | \multicolumn{10}{c}{male tribal populations} | | | | | | | | | |
| ter | KOLAM | | KOYA | | MANNE | | MARIA | | RAJ GOND | |
	γ_1	γ_2	γ_1	γ_2	γ_1	γ_2	γ_1	γ_2	γ_1	γ_2
H.B.	.15	--62	.39	.37	1.62*	4.54*	−.27	.48	−.30	.23
					.71*	.29				
H.L.	−.14	−.06	.48	1.12	−.05	−.08	.05	−.09	−.32	.28
Bg.B.	.83*	2.93*	.17	.19	1.72*	8.42*	−.17	−.63	−.12	−.61
	−.14	−.03			−.40	.27				
T.F.L.	−.26	−.07	.44	.11	.66*	.32	−.05	−.10	−.04	−.24
U.A.L.	−.05	−.63	−1.95*	6.88*	−.01	−.27	.13	.76	.14	−.40
			−.30	.74						
L.A.L.	−2.17*	9.98*	−.07	.59	.19	−.67	−.02	.28	−.06	−.67
	.08	−.62								

The values in the second line for each character are calculated after omitting extreme observations.

Simple graphical displays like histograms and bivariate charts can be of help in detecting outliers and clusters in data. With sophisticated computer graphic facilities now available, the statistician is able to look at many plots during the statistical analysis and thus interact with data in a more effective way. A good reference to graphical techniques is a recent book by Cleveland (1993). In his book on *Statistical Methods for Research Workers*, Fisher (1925) emphasizes the importance of diagrams for preliminary examination of data. With the appearance of Tukey's (1977) pioneering book, *Exploratory Data Analysis*, visualization became far more concrete and effective.

2.3   *Faking of data*

> *The government are very keen on amassing statistics. They collect them, add them, raise them to the n-th power, take the cube root and prepare wonderful diagrams. But you must never forget that every one of these figures comes in the first instance from the village watchman who puts down what he damn pleases.*
> Sir Josiah Stamp (*Playboy Magazine*, Nov.1975)

> *As more cases of fraud broke into public, and whispers were heard of others more quietly disposed of, we wondered if fraud wasn't quite regular minor feature of the scientific landscape.*
> William Broad and Nicholas Wade
> (from *Betrayers of the Truth*)

Since the acceptance of a theory depends on its verification by observed data, a scientist may be tempted to fudge experimental data to fit a particular theory and claim acceptance or establish priority for his ideas. No doubt, if a theory is wrong, it will be discovered sooner or later by other scientists by conducting relevant experiments. However, there is a possibility of considerable harm being done to society by its acceptance in the meanwhile. A recent example is the "IQ Fraud" (*Science Today*, December 1976, p.33) involving Cyril Burt, the undisputed father of British Educational Psychology. His theory that differences in intelligence are largely inherited and not affected by social factors, apparently supported by faked data, influenced government thinking on education of children in the wrong direction.

How can we detect whether given data are faked or not? Does the statistical repertoire include methods of data analysis to indicate if data are not genuine? Fortunately yes. In fact, during recent years statisticians have examined the data sets generated and used by some of the famous scientists in the past and discovered that they "were not all so honest and they did not always obtain the results they reported." Haldane (1948) pointed out:

> Man is an orderly animal. He cannot imitate the disorder of nature.

Based on this limitation of the human brain, statisticians have evolved techniques to detect fakes. The following experiment conducted by me with the students of the first year class in statistics demonstrates Haldane's observation.

I asked the students in one of my classes to do the following experiments, the results of which are given in Table 3.3.

(i) Throw a coin 1000 times and record the number of heads in sets of 5 (column 3, simulated data).

(ii) Find out from a maternity hospital records the number of male children born in 200 sets of 5 consecutive deliveries (column 2, hospital data).

(iii) Imagine that you are throwing a coin and write down the results of 1000 imaginary throws, and find the frequency distribution of the number of heads in sets of 5 throws (column 5, imaginary data A).

(iv) The students had not yet learnt the derivation of the binomial distribution. But I showed them what I expect the frequency distribution of heads in sets of 5 throws to be (column 4 of the table), and asked them to write down the results of 1000 imaginary throws (column 6, imaginary data B).

Table 3.3: Results of different experiments

No of boys	Real data		Expectation (binomial distirbution)	Imaginary data	
(in sets of 5)	hospital	simulated		(A)	(B)
(1)	(2)	(3)	(4)	(5)	(6)
0	2	5	6.25	2	5
1	26	27	31.25	20	32
2	65	64	62.50	78	63
3	64	68	62.50	80	61
4	31	32	31.25	17	33
5	9	4	6.26	3	6
Total	200	200	200.00	200	200
χ^2	2.10	2.18		23.87	0.54

It is seen that the chi-square values, on 5 degrees of freedom each, measuring the deviations from the expected are moderate for the real data. The chi-square value for the imaginary data A is large, since the students imagined more sets balanced for boys and girls than possible by random chance. The chi-square value for the imaginary data B, when the students knew what was expected, is incredibly small showing that they tried to fit the data to known expectations.

Now let us look at the live data from experiments conducted by Mendel, on the basis of which Mendel formulated the laws of heritance of characters and laid the foundations of genetics. In a

**Table 3.4. χ^2 value of deviation from expected and
probability ($\chi^2 >$ observed value) for each group of
experiments conducted by Mendel**

(Source: R.A. Fisher, *Annals of Science*, I, 1936)

Experiments to test hypothesis	degrees of freedom	χ_0^2 (observed)	$P(\chi^2 > \chi_0^2)$
3:1 ratios	7	2.1389	0.95
2:1 ratios	8	5.1733	0.74
bifactorial	8	2.8110	0.94
gametic ratios	15	3.6730	0.9987
trifactorial	26	15.3224	0.95
total	64	29.1186	0.99987
illustrations of plant variation	20	12.4870	0.90
total	84	41.6056	0.99993

remarkable study, R.A. Fisher (*Annals of Science*, 1, 1936, pp.115-137), examined the data by computing the chi-square values measuring the departure from Mendel's theory in groups of experiments. The results are reported in Table 3.4.

We see from the last column of Table 3.4 that the probabilities are extremely high in each case indicating that "data are probably faked to show a remarkably close agreement with theory." The overall probability of such good agreement is

$$1 - .99993 = 7/100000$$

which is very small, Fisher commented on this rare chance as follows:

Although no explanation can be expected to be satisfactory, it remains the possibility among others that Mendel was deceived by some assistant who knew too well what was expected. This possibility is supported by independent evidence that the data of most, if not all, of the experiments have been falsified so as to agree closely with Mendel's expectations.

Haldane (1948) provides several examples of data reported by geneticists which show a high degree of closeness with the postulated theory. Haldane mentions that, if an experimenter knew what tests a statistician would employ to detect faking of data, he might fake in such a way that the data would not look suspicious by these tests and yet would support his theory within the limits of sampling errors. Haldane calls this second order faking. For instance, if theory suggest a 3:1 ratio of two types of events, two numbers could always be chosen such that their ratio is not close to or far from 3:1, so that the chi-square value of deviation from theory is neither too small nor too large. However, there are statistical tests by which such second order faking could be detected.

I have asked one of my colleagues, who is a scientist, to write down an imaginary sequence of fifty H's and T's to support a theory specifying 1:1 ratio for H's and T's but not showing too close an agreement to arouse suspicion. He gave the following sequence which has 29 H's and 21 T's.

```
T  H  T  H  T  H  H  T  H  H
H  T  T  H  T  H  T  H  H  H
T  H  H  H  T  H  T  H  T  T
H  H  T  T  H  T  T  H  H  H
T  H  H  T  T  H  H  H  T  H
```

The chi-square for testing departure from 1:1 ratio is

$$\chi^2 = \frac{(29 - 25)^2}{25} + \frac{(21 - 25)^2}{25} = 1.28$$

which, on one degree of freedom, is neither too small to suggest faking nor too large to reject theory. On the other hand, it is seen that the numbers of H's in the five rows of sequences of ten H's and T's

$$6, 6, 5, 6, 6$$

seem to be more uniform than what is expected by chance. The chi-square for these values is

$$\chi^2 = \frac{2}{5} + \frac{2}{5} + 0 + \frac{2}{5} + \frac{2}{5} = \frac{8}{5} = 1.6$$

on 5 degrees of freedom, which is incredibly small indicating "second order faking."

According to R.S. Westfall (*Science*, 179, 1973, pp.751-758), Newton, the boy genius who formulated the laws of gravitation, was a master at manipulating observations so that they exactly fitted his calculations. He quotes three specific examples from the *Principia*. To establish that acceleration of gravity at the Earth's surface is equal to the centripetal acceleration of the Moon in its orbit, Newton calculated the former as

$$15 \; ft. \; 1 \; inch \; 1 \; \frac{7}{9} \; lines$$

and the latter as

$$15 \; ft. \; 1 \; inch \; 1 \; \frac{1}{2} \; lines$$

respectively, where one line $=1/12$ inch, giving a precision of 1 part in 3000 for comparison. The velocity of sound was estimated to be 1142 ft. per second which has a precision of 1 part in 1000. Newton computed the precision of the equinoxes to be $50^{ii} \; 01^{iii} \; 12^{iv}$, which has a precision of 1 part in 3000. Such a high degree of precision was unheard of with observational techniques in Newton's times.

In the Chapter on Deceit in History in the book *Betrayers of the Truth* by William Broad and Nicholas Wade, the names of other famous scientists who probably faked data are mentioned. I quote:

* Claudius Ptolemy, known as "the greatest astronomer of antiquity" did most of his observing not at night on the coast of Egypt but during the day, in the great library at Alexandria, where he appropriated the work of a Greek astronomer and proceeded to call it his own.

* Galileo Galilei is often hailed as the founder of modern scientific method because of his insistence that experiment, not the works of Aristotle, should be the arbiter of truth. But colleagues of the seventeenth-century Italian physicist had difficulty reproducing his results and doubted if he did certain experiments.

* John Dalton, the great nineteenth-century chemist who discovered the laws of chemical combination and proved the existence of different types of atoms, published elegant results that no present-day chemist has been able to repeat.

* The American physicist Robert Millikan won the Nobel prize for being the first to measure the electric charge of an electron. But Millikan extensively misrepresented his work in order to make his experimental results seem more convincing than in fact was the case.

Why did some of the famous scientists manipulate facts? What could have happened if they were more honest? (These questions were raised by Dr. J.K. Ghosh.)

To answer these questions one must recognize the several facets of a scientific discovery - finding facts (data), postulating a theory or a law to explain the facts and the desire of a scientist to establish priority to gain the respect of one's peers and reap the benefits of recognition. When a scientist was *convinced* of his theory, there was the temptation to look for "facts" or distort facts to fit the theory. The concept of agreement with theory within acceptable margins of error did not exist until the statistical methodology for testing of hypotheses was developed. It was thought that a closer agreement with data implied a more accurate theory and a more convincing evidence for acceptance by the peers. We are now aware - this is due to the emergence of statistical ideas - that too close an agreement with data might imply a spurious theory! In recent times, there have been many instances where data have been faked to establish wrong hypotheses (as in the case of Sir Cyril Burt). They have resulted in considerable harm to society and progress of sciences.

2.4   *Lazzarini and an estimate of* π

In the first Chapter, we saw how Monte-Carlo methods of simulation using random numbers enable us to solve complicated problems which are mathematically intractable such as computing complicated integrals, areas of complex figures, estimation of unknown parameters, etc. I shall give you an interesting application of the Monte-Carlo method for the estimation (obtaining the value) of

$$\pi = 3.14159265...$$

which is the ratio of the perimeter of a circle to its diameter.

Many of you have heard about the Buffon needle problem. In the eighteenth century, the French naturalist Comté de Buffon worked out the probability that a needle of length l thrown at random on to a grid of parallel lines with distance $a(>l)$ apart cuts a line to be $p = 2l/\pi a$. Now, if we conduct an experiment by repeatedly throwing a needle a large number N of times and find that the needle cuts a line R times, then R/N is an estimate of p with the property

$$R/N \to p \quad \text{almost surely, as } N \to \infty$$

i.e., R/N will be invariably close to p as N becomes large. Then a Monte-Carlo estimate of π is obtained from the approximate equation

R/N = 2*l*/πa, giving an approximate value of π (when *l*/*a* is known)
as

$$\hat{\pi} \simeq \frac{2l}{a} \frac{N}{R}. \tag{F}$$

If we did not have any computational method for determining π, we could have estimated it by the formula (F) which needs only a needle of known length *(l)*, a piece of paper with parallel lines drawn on it at a given distance *a* apart, and perhaps good deal of patience in throwing the needle in a mechanical way a large number of times.

Some people had the patience to do this and report the value of π they obtained. Of course, not all experiments would yield the same answer. But if N is large, the different estimates should agree closely. It is on record that a Professor Wolf of Frankfurt threw a needle 5000 times during the decade 1850-60; the needle was 36 mm. long and the plane was ruled 45 mm. apart. He observed that the needle crossed a line 2532 times. An application of the formula (F) gave the estimate, π=3.1416 with an error of 0.6 percent. In the decade 1890-1900, a Captain Fox is stated to have made some 1200 trials "with additional precautions" finding π=3.1419. The most accurate estimate of π was credited to an Italian Mathematician, Lazzarini (often misspelled as Lazzerini by those who referred to his work later). He reported in great detail, in a paper published in the 1901 volume of *Periodico di Matematica*, an experiment based on 3408 trials which resulted in 1808 successes leading to the equation

$$\frac{1808}{3408} = \frac{2l}{\pi a} = \frac{5}{3\pi}$$

using the known ratio *l*/*a*=5/6, giving an estimate of π

$$\hat{\pi}=\frac{5}{6} \cdot \frac{3408}{1808}=\frac{5}{3} \cdot \frac{16 \times 213}{16 \times 113}=\frac{5}{3} \cdot \frac{213}{113}=\frac{355}{113}=3.1413929$$

which differs from the true value only in the 7th decimal place!

Notice the strange numbers that appear in the above computation and how the numbers factorize nicely yielding the value of π as the ratio 355/113 which is known to be the best rational approximation to π involving small numbers (due to the 5th century Chinese Mathematician Tsu Chung-Chin). The next best rational approximation is 52163/16604 involving rather large numbers. The game played by Lazzarini is now clear as revealed by independent investigations due to N.T. Gridgman *(Scripta Mathematica,* 1961) and T.H. O'Beirne *(The New Scientist,* 1961, p.598). In order to get the ratio 355/113 when $l/a = 5/6$, one has to get the ratio 113/213 for R/N, i.e., 113 successes in 213 trials (at the minimum) or 113 k successes in 213 k trials for any integer k. In Lazzarini's case k was 16. There are two possibilities. Either he did not do any experiments which he described in great detail in his article and just reported the numbers he wanted. Or, he did experiments in batches of 213 trials and "watched his step" till he struck the right number of successes. With 16 repetitions, as done by Lazzarini, the chance of getting the right number of successes, 113x16, is about 1/3.

Laplace, in his *Theórie Analytique des Probabilités* wrote:

> *It is remarkable that a science which began with consideration of games of chance should have become the most important object of human knowledge.*

He did not envisage that a technique used to acquire new knowledge could be manipulated to support a wrong claim. Laplace must have thought that such frauds would be discovered sooner or later, perhaps, through considerations of the same games of chance.

2.5   *Rejection of outliers and selective use of data*

Charles Babbage, the inventor of a calculating machine that was the forerunner of the computer, in his book *Reflections on the*

Decline of Science in England, written in 1830, categorized different types of non-cavalier attitudes to data and its use by the scientist.

(i)   *Trimming:* "Clipping off little bits here and there from those observations which differ most in excess from the mean, and in sticking them onto those which are too small."

(ii)   *Cooking:* "Art of various forms, the object of which is to give ordinary observations the appearance and character of those of the highest degree of accuracy. One of its numerous processes is to make multitudes of observations, and out of these to select only those which agree, or very nearly agree. If a hundred observations are made, the cook must be very unhappy if he cannot pick out fifteen or twenty which will do for serving up."

(iii)   *Forging:* "Recording of observations never made."

I have already discussed about forging or producing data out of thin air. I shall now discuss the more troublesome problem of dealing with outliers and other inconsistencies in data.

How do we deal with observations which look extreme or, in some way, inconsistent with the others? This perplexing problem described as that of "outliers" and "contamination" is one of the modern areas of research. Unfortunately, no satisfactory solution has been put forward, except rationalizing and making some statistical adjustments for trimming. Perhaps, a more scientific approach when outliers are suspected is to consider the following possibilities.

*   An outlier may be the result of a gross error in measurement or recording.

*   The unit (or individual) associated with the outlier does not belong to the population under study, or is distinguishable in some qualitative way from the others

in the sample.
*   The population under study has a heavy tailed distribution so that the occurrence of large values is not rare.

The first step in dealing with what appears to be outliers is to identify the relevant units in the population if possible and review each case in the light of the alternatives listed above. It may be possible, to find a suitable explanation suggesting appropriate action to be taken. Occasionally, re-examination of an aberrant measurement may lead to a new discovery! Such an investigation, going back to the source of measurements, may not always be possible, which underlines the importance of incorporating automatic scrutiny of data while collecting, and recording supplementary information when a measurement is suspected to be an outlier. When re-examination of sampled units is not possible or expensive, one may have to depend on purely statistical tests to decide whether:

*   To reject outlying observations and treat the rest as a regular (valid) sample from the population under study.
*   To reject outlying observations and make adjustments for them in statistical analysis.
*   To accept ("it would be more philosophical") what seemed to be outliers as a normal phenomenon of the population under study and use an appropriate model for statistical analysis.

The present statistical methodology is not adequate to deal with the problems outlined above, but the different directions in which statisticians are currently working, such as robust inference, detection of outliers and influential observations, may provide a unified theory for incorporating the information acquired through cross examination of data in inferential data analysis. However, I shall leave one thought to the reader.

To omit or not to omit an outlier or a spurious observation is a serious dilemma as the following example shows. Suppose that we have N observations from a population with mean μ and standard deviation (s.d.) σ giving a mean value $\bar{x}$, and M spurious observations from another population with mean ν and s.d. σ giving a mean value $\bar{y}$. Let us ignore the fact that $\bar{y}$ arose from contaminating observations and estimate μ by

$$\hat{\mu} = \frac{(N\bar{x} + M\bar{y})}{(N + M)}.$$

Then, denoting $\nu - \mu = \delta\sigma$,

$$E(\hat{\mu} - \mu)^2 = \frac{\sigma^2}{N + M} \left[1 + \frac{M^2\delta^2}{N + M} \right] < V(\bar{x}) = \frac{\sigma^2}{N}$$

if $\delta^2 < M^{-1} + N^{-1}$ which is always true when $\delta \leq 1$ and $M = 1$ whatever N may be. Thus under the mean squared error criterion, which is popular among statisticians, it pays to include a spurious observation from a population whose mean may differ by as much as one standard deviation from the parameter under estimation! Such an improvement can be of considerable magnitude in small samples.

3. Meta analysis

Teacher: *Which is more important, the Sun or the Moon?*
Student: *Of course, the Moon as it gives light when it is badly needed!*

In making decisions, one has to take into account all the available evidence which may be in the form of several pieces of information gathered from different sources, some of which may be in the form of expert opinions. Several questions arise in this connection.

* How reliable is each piece of information?
* How much of this information is relevant to the problem under investigation?
* Are different pieces of information consistent?
* How do we pool the information from different sources which may not all be consistent, to arrive at a conclusion?

These are not new questions individually but their collective consideration in an investigation is not usually emphasized. Attempts are being made to lay down systematic procedures to study these questions under the title, meta analysis.

The major sources of information on any problem are papers published in journals or in special reports. But those may not represent all the studies done on a given problem. For instance, studies which do not yield successful results do not get published. Editors of journals discourage publication of studies which do not yield results that are statistically significant at the traditional levels (e.g., $p < .05$). Such unpublished papers end up in the *file drawers* of the investigators and not are available for review. In meta analysis, the bias arising out of excluding unfavorable studies is referred to as the *file drawer problem*. Some methods have been suggested for making adjustments to minimize the effect of such a bias.

Evaluation of each piece of information enables us to determine the weight to be attached to it in pooling information. However, pooling demands that different pieces of information are not conflicting with each other. Finally, a choice has to be made of an appropriate method to combine the different pieces of information and express the reliability of the final conclusion. All these require a judicious use of the whole battery of available statistical methodology from scrutiny of data to inferential data analysis, and perhaps, a philosophical approach to problem solving as indicated by the dialogue between the teacher and the student quoted above.

4.   Inferential data analysis and concluding remarks

It is an extraordinary thing, of course, that everybody is answering questions without knowing what the questions are. In other words, everybody is finding some remedy without knowing what the malady is.

Jawaharlal Nehru

Inferential data analysis refers to the statistical methodology based on a specified underlying stochastic model, for estimating unknown parameters, testing specified hypotheses, prediction of future observations, making decisions, etc. The choice of a model may depend on the specific information we are seeking from data. It may not necessarily be the one which explains the whole observed data, but one which provides efficient answers to specified questions.

Data analysis for answering specific questions raised by customers is not the only task of a statistician. A wider analysis for understanding the nature of given data would be of use in finding which questions can be answered with available data, in rising new questions and in planning further investigations.

It is also a good practice to analyze given data under different alternative stochastic models and to examine differences in conclusions that emerge. Such a procedure may be more illuminating than seeking for robust inference procedures to safeguard against a wide class alternative stochastic models. The possibility of using different models for the same data to answer different questions should also be explored.

Inferential data analysis should be of an interactive type: new features of the data may emerge during the analysis under a specified model requiring a change in the analysis originally contemplated.

Simulation studies to assess the performance of certain procedures and bootstrap and jack-knife techniques for estimating variances of estimators (Efron, (1979)) under complicated data structures, which depend on the heavy use of computers, have given additional dimensions to data analysis, although some caution is needed in interpreting the results of such analyses.

There is the usual dictum in inferential data analysis that once the validity of a model is assured, there is an optimum way of analyzing the data such as the use of $\bar{x}$ as an estimate of the mean of a normal population based on a given sample, or of the mean of a finite population based on a random sample without replacement. As an example of the latter case, suppose that the problem is that of estimating the average yield of trees planted in a row by taking a sample of size 3. Our prescription says that if x_1, x_2, x_3 are the observed yields on three randomly chosen trees, then a good estimate is $\bar{x} = (x_1 + x_2 + x)/3$. However, if after drawing the sample we find that two of the trees chosen are next to each other with the corresponding yields, say x_1 and x_2, then we may be better off in giving the alternative estimator $\hat{x} = (y + x_3)/2$ where $y = (x_1 + x_2)/2$. It may be seen that if the yields of consecutive trees are highly correlated, then the variance of $\hat{x}$ is less than that of $\bar{x}$ in samples where at least two consecutive trees are chosen. Such strategies as using different methods for different configurations of the sample under the same stochastic model should be explored.

Then, there is the problem of "Oh! Calcutta." Suppose that someone who is not aware of the large differences in the populations of Calcutta and the rest of the towns and cities (which we refer to as units) in the state of West Bengal tries to estimate the total population of the state by taking a simple random sample of the units without replacement. The usual formula in such a case, which is proved to be optimal in many ways, is $N\bar{x}$, where N is the total number of units

in West Bengal and $\bar{x}$ is the average population in the sample of n

randomly chosen units. Let us suppose that Calcutta comes into the sample, whose population is several times that of any other unit in

West Bengal. In such a case it would be disastrous to suggest $N\bar{x}$

as the estimate of the total population, especially when n, the sample size, is small. Suppose x_1 in the sample is the population of Calcutta, then a reasonable estimate of the total population of West Bengal would be

$$x_1 + \frac{N - 1}{n - 1} (x_2 + \dots + x_n) .$$

What we have done is post stratification after looking at a particular observed data set!

As statisticians, we are asked to advise on the appropriate statistical methodology (or software package program) for a certain data set without having the opportunity to cross examine the data. Our answer should be: statistical treatment cannot be prescribed over the phone or bought over the counter. The data has to be subjected to certain diagnostic tests and special features, if any, have to be taken into account, and then a course of treatment is prescribed and the progress is continuously monitored to decide on any changes needed in the treatment.

Let me conclude with the following summary. The purpose of statistical analysis is "to extract all the information from observed data." The recorded data may have some defects such as recording errors and outliers or may be faked, and the first task of a statistician is to scrutinize or cross examine the data for possible defects and understand its special features. The next step is the specification of a suitable stochastic model for the data using prior information and cross-validation techniques. On the basis of a chosen model, inferential analysis is made, which comprises of estimation of unknown parameters, tests of hypotheses, prediction of future observations and decision making. Examining data under different possible models is suggested as more informative than using robust procedures to safeguard against possible alternative models. Data analysis must also provide information for raising new questions and for planning future investigations.

Finally, I must stress the need for active collaboration between statisticians and experimental scientists. A statistician can help the scientist in designing efficient experiments to yield the maximum information on the questions raised by the scientist and providing the scientist guidelines for examining his hypotheses and modifying them if the data indicate contrary evidence. As Fisher, the father of modern experimental designs said:

> To consult a statistician after an experiment is finished is often merely to ask him to conduct a post mortem examination. He can perhaps say what the experiment died of.

References

Chatfield, C. (1985). The Initial examination of data. *J. Roy. Stat. Sco.* A, **148**, 214-253.

Cleveland, W.S. (1993). *Visualizing Data*, AT&T Bell Laboratories, Murray Hill, New Jersey.

Efron, B. (1979). Bootstrap methods: Another look at jack-knife. *Ann. Statist.* **7**, 1-26.

Fisher, R.A. (1922). On the mathematical foundations of theoretical statistics. *Philos. Trans. Roy. Soc.* **222**, 309-368.

Fisher, R.A. (1925). *Statistical Methods for Research Workers*, Olivia and Boyd.

Fisher, R.A. (1934). The effect of method of ascertainment upon estimation of frequencies. *Ann. Eugen.* **6**, 13-25.

Fisher, R.A. (1936). Has Mendel's work been rediscovered? *Annals of Science* **1**, 115-137.

Fox, J.P., Hall, C.E. and Elveback, L.R. (1970). *Epidemiology, Man and Disease*, MacMillan Co, London.

Hacking, Ian (1984). Trial by number. *Science* **84**, 69-70.

Haldane, J.B.S. (1948). The faking of genetic results. *Eureka* **6**, 21-28.

Mahalanobis, P.C. (1931). Revision of Risley's anthropometric data relating to the tribes and castes of Bengal. *Sankhyā* **1**, 76-105.

Mahalanobis, P.C. (1944). On large scale sample surveys. *Philos. Trans. Roy. Soc.*, London, Series B, **231**, 329-451.

Majumdar, D.N. and Rao, C. Radhakrishna (1958). Bengal anthropometric survey, 1945: A statistical study. *Sankhyā,* **19**, 201-408.

Mosteller, F. and Tukey, J.W. (1968). Data analysis including statistics. In *Handbook of Social Psychology*, Vol. **2** (Eds. G. Linzey and E. Aronson), Addison-Wesley.

Mukherji, R.K., Rao, C.R. and Trevor, J.C. (1955). *The Ancient Inhabitants of Jebel Moya.* Cambridge University Press.

Neyman, J. and Pearson, E.S. (1966). *Joint Statistical Papers by J. Neyman and E.S. Pearson*, Univ. of California Press, Berkeley.

Pearson, K. (1914-15). On the probability that two independent distributions of frequency are really samples of the same population, with special reference to recent work on the identity of Trypanosome strains. *Biometrika*, **10**, 85-154.

Pitman, E.J.G. (1937). Significance tests which may be applied to samples from any population. *J. Roy. Statist. Soc.* Ser. B, **4**, 119-130.

Rao, C. Radhakrishna (1948). The utilization of multiple measurements in problems of biological classification. *J. Roy. Statist. Soc.* B, **10**, 159-203.

Rao, C. Radhakrishna (1971). Taxonomy in anthropology. In *Mathematics in Archeological and Historical Sciences,* Edin. Univ. Press, 329-358.

Rao, C. Radhakrishna (1987). Prediction of future observations in growth curve models. *Statistical Sciences*, **2**, 434-471.

Shewart, W.A. (1931). *Economic Control of Quality of Manufactured Product,* D. Van Nostrand, New York.

Tukey, J. (1962). The future of data analysis. *Ann. Math. Statist.*, **30**, 1-67.

Tukey, J. (1977). *Exploratory Data Analysis* (EDA), Addison Wesley.

Urmila Pingle (1982). Morphological and Genetic Composition of Gonds of Central India: A statistical study, Ph.D. Thesis, Submitted to Indian Statistical Institute.

Wald, A. (1950). *Statistical Decision Functions,* Wiley, New York.

Additional References Not Cited in Text

Andrews, D.F. (1978). Data analysis, exploratory. In *International Encyclopedia of Statistics* (W.H. Kruskal and J.M. Tanur, ed.), 97-106. The Free Press, New York.

Anscombe, F.J. and Tukey, J.W. (1963). The examination and analysis of residuals. *Technometrics, 5*, 141-160.

Bertin, J. (1980). *Graphics and Graphical Analysis of Data.* DeGruyter, Berlin.

Mallows, C.L. and Tukey, J.W. (1982). An overview of the techniques of data analysis, emphasizing its exploratory aspects. In *Some Recent Advances in Statistics*, 113-172, Academic Press.

Rao, C.R. (1971). Data, analysis and statistical thinking. In *Economic and Social Development, Essays in Honor of C.D. Deskmukh,* 383-392 (Vora and Company).

Solomon, H. (1982). Measurement and burden of evidence. In *Some Recent Advances in Statistics*, 1-22, Academic Press.

Watcher, K.W. and Straff, M.L. (1990). *The Future of Meta Analysis*, Russel Sage Foundation.

Chapter 4

Weighted Distributions - Data with Built-in Bias

The sciences do not try to explain, they hardly even try to interpret, they mainly make models. By a model is meant a mathematical construct which, with the addition of certain verbal interpretations, describes observed phenomena. The justification of such a mathematical construct is solely and precisely that it is expected to work.

von Neumann

1. Specification

In statistical inference, i.e., making statements about a population on the basis of a sample drawn from it, it is necessary to identify the set of all possible samples that could be drawn, designated by Ω and the family of probability distributions to which the actual probability distribution governing the samples belong, designated by P. Much depends in inferential analysis on the choice of P called specification. Wrong specification may lead to wrong inference, which is sometimes called the *third kind of error* in statistical parlance.

The problem of specification is not a simple one. A detailed knowledge of the procedure actually employed in acquiring data is an essential ingredient in arriving at a proper specification. The situation is more complicated with field observations and nonexperimental data, where nature produces events according to a certain stochastic model, and the events are observed and recorded by field investigators. There does not always exist a suitable sampling frame for designing a sample survey to ensure that the events which occur have specified (usually equal) chances of coming into the sample. In practice, all the events that occur in nature cannot be brought into the sample frame.

95

For instance, certain events may not be observable and therefore missed in the record. This gives rise to what are called *truncated, censored or incomplete samples*. Or, an event that has occurred may be observable only with a certain probability depending on the nature of the event, such as its conspicuousness and the procedure employed to observe it, resulting in *unequal probability sampling*. Or, an event which has occurred may change in a random way by the time or during the process of observation so that what comes on record is a modified event, in which case the *change* or *damage* has to be appropriately modeled for statistical analysis. Sometimes, events from two or more different sources having different stochastic mechanisms may get mixed up and brought into the same record, resulting in *contaminated* samples. In all of these cases, the specification for the original events (as they occur) may not be appropriate for the events as they are ascertained (observed data) unless it is suitably modified.

In a classical paper, Fisher (1934) demonstrated the need for an adjustment in specification depending on the way data are ascertained. The author extended the basic ideas of Fisher in Rao (1965, 1973, 1975, 1977, 1985) and developed the theory of what are called weighted distributions as a method of adjustment applicable to many situations. We discuss some applications of weighted distributions outlining the general theory. This Chapter can be read skipping the demonstration of some mathematical results.

2.   Truncation

Some events, although they occur, may be unascertainable, so that the observed distribution is truncated to a certain region of the sample space. For instance, if we are investigating the distribution of the number of eggs laid by an insect, the frequency of *zero eggs* is not ascertainable. Another example is the frequency of families where both parents are heterozygous for albinism but have no albino children. There is no evidence that the parents are heterozygous unless they have an albino child, and the families with such parents and having no albino children get confounded with normal families. The actual frequency of the event *zero albino children* is thus not ascertainable.

In general, if p(x,θ) is the p.d.f. (probability density function for a continuous variable or probability for a discrete variable), where θ denotes an unknown parameter, and the random variable X is truncated to a specified region $T \subset \Omega$ of the sample space, then the p.d.f. of the truncated random variable X^T is

$$p^T (x, \theta) = \frac{w(x, T) \, p(x, \theta)}{u(T, \theta)} \qquad (2.1)$$

where w(x,T)=1 if $x \in T$ and = 0 if $x \notin T$, and u(T,θ)=E[w(X,T)]. The expression (2.1) is the original probability density weighted by a suitable function, and it provides a simple example of a weighted probability distribution whose general definition is given in the next section.

Suppose the event zero is not observable in sampling from a binomial distribution with index n and probability of success π. Let R^T denote the TB (truncated binomial) random variable. Then

$$P(R^T=r)=\frac{n!}{r!(n-r)!} \frac{\pi^r(1-\pi)^{n-r}}{1-(1-\pi)^n}, \quad r=1,\ldots,n. \qquad (2.2)$$

For such a distribution

$$E(R^T)=\frac{n\pi}{1-(1-\pi)^n}, \quad E(R^T/n)=\frac{\pi}{1-(1-\pi)^n} \qquad (2.3)$$

which are somewhat larger than those for a complete binomial, for which the above values are nπ and π respectively.

The following data relate to the numbers of brothers and sisters in families of the girls whose names were found in a private telephone notebook of a European professor. (The first number within the brackets gives the numbers of sisters including the respondent and the second number, that of her brothers.)

$$(1,0),(1,0),(1,1),(1,1),(1,1),(1,1),(1,1),(1,1),(1,1),(1,1)$$
$$(1,1),(2,0),(2,0),(2,0),(2,1),(2,1),(2,1),(2,1),(1,2),(1,2)$$
$$(3,0),(3,1),(3,1),(1,3),(1,3),(4,0),(4,1),(1,4). \qquad (2.4)$$

Since at least one girl is present in the family, we may try and see whether the data conform to a TB distribution with the observation on *zero sisters* missing (i.e., Binomial truncated at zero). The expected number of girls under this hypothesis, assuming $\pi=0.5$, is

$$\sum_{n-1}^{5} f(n) \, E(r|n) \qquad (2.5)$$

where f(n) is the observed number of families with size n(i.e., the total number of brothers and sisters). Using the formulas (2.3) and (2.5) and the data (2.4), we have:

Number of	observed	expected
Sisters	47	46
Brothers	30	31

The observed figures seem to be in good agreement with those expected under the hypothesis of truncated binomial. However, a different story may emerge in a similar situation as in the following data giving the numbers of sisters and brothers in the families of girl acquaintances of a male student in Calcutta.

$$(2,1),(1,1),(3,0),(2,0),(3,1),(1,0),2,1),(1,0),(1,1),(1,1). \quad (2.6)$$

The expected numbers of sisters under the hypothesis of truncated binomial is 14.6 (using the formulas (2.3) and (2.5)) whereas the observed number is 17. The truncated binomial is not appropriate for the data (2.6) and it appears that the mechanisms of *encountering girls* seem to be different in the cases of the European professor and the Calcutta student.

Note that if we sample a number of households in a city and ascertain the numbers of brothers and sisters (i.e., sons and daughters) in each household, then we expect the number of sisters to follow a complete binomial distribution. If from such data we omit the households which do not have girls, then the data would follow a truncated binomial distribution. The professor seems to be sampling from the general population of households with at least one girl. We shall see in the next section that a different distribution holds when data are ascertained about sisters and brothers from boys or girls one *encounters*. The case of the student seems to fall in such a category.

3.   Weighted distributions

In Section 2, we have considered a situation where certain events are unobservable. But a more general case is when an event that occurs has a certain probability of being recorded (or included in the sample). Let X be a random variable with $p(x,\theta)$ as the p.d.f., where θ is a parameter, and suppose that when X=x occurs, the probability of recording it is $w(x,\alpha)$ depending on the observed x and possibly also on an unknown parameter α. Then the p.d.f. of the resulting random variable X^w is

$$p^{w}(x,\theta,\alpha)=\frac{w(x,\alpha)p(x,\theta)}{E[w(X,\alpha)]}. \qquad (3.1)$$

Although in deriving (3.1), we chose $w(x,\alpha)$ such that $0\leq w(x,\alpha)\leq 1$, we may formally define (3.1) for any arbitrary nonnegative function $w(x,\alpha)$ for which $E[w(X,\alpha)]$ exists. The p.d.f. so obtained is called a weighted version of $p(x,\theta)$ and denoted by $p^w(x,\theta)$. In particular the weighted distribution

$$p^w(x,\theta) = \frac{f(x)p(x,\theta)}{E(f(X))} \qquad (3.2)$$

where f(x) is some monotonic function of x, is called a size biased distribution. When X is univariate and nonnegative, the weighted distribution

$$p^w(x,\theta,) = \frac{x^\alpha p(x,\theta)}{E(X^\alpha)} \qquad (3.3)$$

introduced in Rao (1965) has found applications in many practical problems (see Rao, 1985)). When $\alpha=1$, it is called length (size) biased distribution. For example, if X has the logarithmic series distribution

$$P(X=r) = \frac{\theta^r}{-r\log(1-\theta)}, \quad r=1,2, \ldots \qquad (3.4)$$

then the distribution of the length biased variable is

$$P(X^w=r) = (1-\theta)\theta^{r-1}, r=1,2,\ldots \qquad (3.5)$$

which shows that X^w-1 has a geometric distribution. A truncated geometric distribution is sometimes found to provide a good fit to an observed distribution of family size (Feller, 1968). But, if the information on family size has been ascertained from school children, then the observations may have a size biased distribution. In such a case, a good fit of the geometric distribution to the observed family size would indicate that the underlying distribution, is, in fact, a logarithmic series.

   In the case of many discrete distributions, as shown in Rao (1965, 1985), the size biased form belongs to the same family as the original distribution. An exception is the logarithmic series distributions.

An extensive literature on weighted distributions has appeared since the concept was formalized in Rao (1965); it is reviewed with a large number of references in a paper by Patil (1984) with special reference to the earlier contributions by Patil and Rao (1977, 1978) and Patil and Ord (1976). Rao (1985) contains an updated review of the previous work and some new results.

4.   P.p.s. sampling

An example of a weighted distribution arises in sample surveys when unequal probability sampling or what is known as p.p.s. (probability proportional to size) sampling is employed. A general version of the sampling scheme involves two random variables X and Y with p.d.f. $p(x,y,\theta)$ and a weight function $w(y)$ which is a function of y only, giving a weighted p.d.f.

$$p^w(x,y,\theta) = \frac{w(y)p(x,y,\theta)}{E[w(Y)]}. \tag{4.1}$$

In sample surveys, we obtain observations on (X^w, Y^w) from the p.d.f. (4.1) and draw inference on the parameter θ.

It is of interest to note that the marginal p.d.f. of X^w is

$$p^w(x,\theta) = \frac{w(x,\theta)p(x,\theta)}{E[w(X,\theta)]} \tag{4.2}$$

which is a weighted version of $p(x,\theta)$ with the weight function

$$w(x,\theta) = \int p(y|x,\theta)w(y)dy. \tag{4.3}$$

If we have a sample of size *n*

$$(x_1,y_1), \ldots, (x_n,y_n) \tag{4.4}$$

from the distribution (4.1), then an estimate of E(X), the mean with respect to the original p.d.f. $p(x,y,\theta)$, which is the parameter of interest, is

$$\frac{E\,[w\,(Y)]}{n} \sum_{i=1}^{n} \frac{x_i}{w\,(y_i)} \tag{4.5}$$

which is an unbiased estimator of E(X). The estimator

$$\frac{1}{n} \sum_{i=1}^{n} x_i \tag{4.6}$$

would be an unbiased estimator of $E(X^w)$, the mean with respect to the weighted p.d.f. $p^w(x,\theta)$ as in (4.2).

5. Weighted binomial distribution: Empirical theorems

Suppose that we ascertain from each *male* member, in a class or in any congregation at any time and at any place, the number of brothers including himself and the number of sisters he has, and raise the following question. What is the approximate value of B/(B+S), where B and S are the total numbers of brothers and sisters in all the families of the male respondents? It is clear that we are sampling from a truncated distribution of families with at least one male member so that B/(B+S) should be larger than one-half. But by how much? Surprisingly, when k, the number of males asked, is not very small, one can make accurate predictions of the relative magnitudes of B and S, and of the ratio B/(B+S). This may be stated in the form of an empirical theorem.

Empirical Theorem 1: *Let k male respondents observed in any gathering **anywhere and at any time** have a total number B of brothers (including themselves) and a total number S of sisters. Then the following predictions hold:*

(i)   B is much greater than S.
(ii)   B-k is approximately equal to S.
(iii)   B/(B+S) is larger than one-half. It will be closer to

$$\frac{1}{2} + \frac{k}{2 \, (B + S)}$$

(iv)   (B-k)/(B+S-k) is close to half.

The roles of B and S are reversed if the data are ascertained from the *female* members in a gathering.

Consider a family with n children. Then on the assumption of a binomial distribution with $\pi = 1/2$ and index n, the probability of r male children is

$$p(r) = \frac{n! 2^{-n}}{r!(n-r)!}, \quad r = 0, 1, 2, \ldots \quad . \tag{5.1}$$

In our case, there is at least one male child so that the appropriate distribution is a truncated one. One possibility is a truncated binomial (TB),

$$p^{T}(r) = \frac{n!}{r!(n-r)!} \, \frac{1}{2^n - 1}, \quad r = 1, 2, \ldots \tag{5.2}$$

and another is a size biased binomial (WB)

$$p^{w}(r) = \frac{2}{n} r \binom{n}{r} \left(\frac{1}{2}\right)^n = \binom{n-1}{r-1} \left(\frac{1}{2}\right)^{n-1}, \quad r = 1, 2, \ldots \quad . \tag{5.3}$$

In Rao (1977), it was argued that (5.3) is more appropriate for the observed data than (5.2). Table 4.1 gives the observed frequency distributions of the number of brothers in families of different sizes based on the data collected separately from the male and female students in the universities at Shanghai (China), Manila (Philippines), and Bombay (India), and the expected values on the hypotheses of TB as in (5.2) and WB as in (5.3).

It is seen from Table 4.1 that the WB (weighted binomial) provides a better fit than the TB (truncated binomial) indicating that a family with r brothers is sampled with probability proportional to r.

Accepting the hypothesis of the weighted (size biased) binomial as in (5.3) we immediately find that

$$E(r \mid n) = \sum_{r=1}^{n} r \binom{n-1}{r-1} \left(\frac{1}{2}\right)^{n-1} = \frac{n+1}{2} \qquad (5.4)$$

$$\Rightarrow E(r-1) = \frac{n-1}{2}. \qquad (5.5)$$

If $(r_1, n_1), \ldots, (r_k, n_k)$ are observed data with $S = T-B$, $B = r_1 + \ldots + r_k$, $T = n_1 + \ldots + n_k$, then for given T

$$E(B-k) = \sum_{1}^{k} E(r_i - 1) = \sum_{1}^{k} \frac{n_i - 1}{2} = \frac{T-k}{2} = E(S). \qquad (5.6)$$

$$E(B) = \frac{T+k}{2}, E\left(\frac{B}{T}\right) = E\left(\frac{B}{B+S}\right) = \frac{1}{2} + \frac{k}{2(B+S)}. \qquad (5.7)$$

Removing the expectation signs in (5.6) and (5.7), we can assert approximate equalities as stated in the Empirical Theorem 1.

During the last twenty years, while lecturing to students and teachers in different parts of the world, I collected data on numbers of brothers and sisters in each family of the members in my audience. The results are summarized in Tables 4.2 - 4.5. It is seen that the predictions as given in the empirical Theorem 1, based on the

Table 4.1 Observed (obs) frequencies of the number of brothers in families of different sizes and expected frequencies under the hypotheses of TB and WB distributions

No. of brothers	n = 1 observed	expected TB	WB	n = 2 observed	expected TB	WB	n = 3 observed	expected TB	WB
1	6	6	6	24	28.7	21.5	12	20.1	11.7
2				19	14.3	21.5	24	20.2	23.6
3							11	6.7	11.7
Total	6	6	6	43	43.0	43.0	47	47.0	47.0

No. of brothers	n = 4 observed	expected TB	WB	n = 5 observed	expected TB	WB	n = 6 observed	expected TB	WB
1	8	11.2	5.3	5	6.5	2.5	1	1.9	0.6
2	10	16.8	15.7	8	12.9	10.0	4	4.8	3.1
3	17	11.2	15.7	15	12.9	15.0	4	6.3	6.3
4	7	2.8	5.3	10	6.5	10.0	9	4.8	6.3
5				2	1.2	2.5	2	1.9	3.1
6							0	0.3	0.6
Total	42	42.0	42.0	40	40.0	40.0	20	20.0	20.0

(Data from male students in Shanghai, Manila and Bombay)

hypothesis of the weighted binomial, are true in practically every case. As a further test of the weighted binomial, the statistic

$$\chi^2 = \frac{4([B-k]-[(T-k)/2])^2}{T-k},$$ (5.8)

which is asymptotically distributed as Chi-square on one degree of freedom, is calculated in each case. The Chi-squares are all small providing evidence in favor of the weighted binomial distribution.

Table 4.2 Data on male respondents (students)

Place and year	k	B	S	$\frac{B}{B+S}$	$\frac{B-k}{B+S-k}$	χ^2
Bangalore(India,75)	55	180	127	.586	.496	.02
Delhi (India,75)	29	92	66	.582	.490	.07
Calcutta (India,63)	104	414	312	.570	.498	.04
Waltair (India,69)	39	123	88	.583	.491	.09
Ahmedabad (India, 75)	29	84	49	.632	.523	.35
Tirupati(India,75)	592	1902	1274	.599	.484	.50
Poona (India.75)	47	125	65	.658	.545	1.18
Hyderabad (India, 74)	25	72	53	.576	.470	.36
Tehran (Iran,75)	21	65	40	.619	.500	.19
Isphahan(Iran,75)	11	43	32	.584	.515	.06
Tokyo (Japan,75)	50	90	34	.725	.540	.49
Lima (Peru,82)	38	132	87	.603	.519	.27
Shanghai(China,82)	74	193	132	.594	.474	.67
Columbus (USA,75)	29	65	52	.556	.409	2.91
College St. (USA, 76)	63	152	90	.628	.497	.01
Total	1206	3734	2501	.600	.503	0.14

k=number of students, B=total number of brothers including respondent, S=total number of sisters. Estimate of π under size biased distribution=(B-k)/(B+S-k).
[Actually, the Chi-squares are too small which needs further study of the mechanism underlying the observed data.]

Table 4.3 Data on female respondents (students)

Place and year	k	B	S	$\dfrac{B}{B+S}$	$\dfrac{B-k}{B+S-k}$	χ^2
Lima (Peru,82)	16	37	48	.565	.464	.36
Los Banos (Philippines,83)	44	101	139	.579	.485	.18
Manila (Philippines,83)	84	197	281	.588	.500	.00
Bilbao (Spain,83)	14	19	35	.576	.525	.10
Shanghai(China,82)	27	28	55	.662	.500	.00

Table 4.4 Data on male respondents (professors)

Place and year	k	B	S	$\dfrac{B}{B+S}$	$\dfrac{B-k}{B+S-k}$	χ^2
State College (USA, 75)	28	80	37	.690	.584	2.53
Warsaw (Poland, 75)	18	41	21	.660	.525	2.52
Poznan (Poland, 75)	24	50	17	.746	.567	1.88
Pittsburgh (USA, 81)	69	169	77	.687	.565	2.99
Tirupati (India,76)	50	172	132	.566	.480	0.39
Maracaibo (Venezuela, 82)	24	95	56	.629	.559	1.77
Richmond (USA,81)	26	57	29	.663	.517	0.03
Total	239	664	369	.642	.535	3.95

Note 1. From (5.7), the expected value of the ratio $B/(B+S)$ for given average family size $f=(B+S)/k$ is as follows for different values of f:

f:	1	2	3	4	5	6
$E[B/(B+S)]$:	1	.75	.67	.625	.6	.58

The situation is slightly different in Table 4.5 relating to the data on professors. The estimated proportion is more than half in each case, and the Chi-square values are high; this implies that the weight function appropriate to these data is of a higher order than r, the number of brothers. Male professors seem to come from families where sons are disproportionately more than the daughters!

These figures show that in any given situation where the average family size is not likely to exceed 6, the following predictions can be made about the total number of brothers (B) and of sisters (S) ascertained from the male members in any gathering:

(i)   B is much greater than S.

(ii)   B/(B+S) is closer to 0.6 or even 2/3 rather than to 1/2.

(iii)   B/(B+S-k) is closer to 1/2 where *k* is the number of males responding to the question.

Surprisingly, these predictions hold even if k, the number of males in a gathering, is small. [This will be a good classroom exercise or a demonstration piece in any gathering. One can make these predictions in advance and demonstrate the accuracy of predictions after collecting the data from male (or female) members.]

Note 2. The probabilities for B>S, B=S, B<S in the case of a weighted binomial distribution for n=1,2, ... are given in Table 4.5. It is seen that P(B>S) is much larger than P(B<S) for each n so that in any given audience, the ratio of b_g (males belonging to families with B>S) to b_l (those with B<S) is likely to be large, depending on the distribution of family sizes.

Table 4.5 Probabilities of B>S, B=S and B<S
(Weighted binomial with weights proportional to number of brothers)

n	1	2	3	4	5	6	7	8	9	10
B>S	1	$\frac{1}{2}$	$\frac{3}{4}$	$\frac{1}{2}$	$\frac{11}{16}$	$\frac{1}{2}$	$\frac{42}{64}$	$\frac{1}{2}$	$\frac{163}{256}$	$\frac{1}{2}$
B=S	0	$\frac{1}{2}$	0	$\frac{3}{8}$	0	$\frac{10}{32}$	0	$\frac{35}{128}$	0	$\frac{90}{512}$
B<S	0	0	$\frac{1}{4}$	$\frac{1}{8}$	$\frac{5}{16}$	$\frac{6}{32}$	$\frac{22}{64}$	$\frac{29}{128}$	$\frac{93}{256}$	$\frac{166}{512}$

We may now state another empirical theorem.

Empirical Theorem 2. *The numbers b_g and b_l are approximately in the ratio of the right hand side expressions in (5.9) and (5.10):*

$$E(b_g) = p_1 + \frac{3}{4}p_3 + \frac{11}{16}p_5 + \ldots + \frac{1}{2}(p_2 + p_4 + \ldots) \qquad (5.9)$$

$$E(b_l) = \frac{1}{4}p_3 + \frac{1}{8}p_4 + \ldots \qquad (5.10)$$

where p_n is the number of families with n children.

In western audiences where the expected family size is small, the ratio $b_g : b_l$ is likely to be even larger than 4:1 and in oriental audiences larger than 2:1, which are quite high compared to 1:1. [This phenomenon can be predicted and verified by asking the members of an audience by show of hands how many belong to the category B>S and how many to B<S. This will be a good classroom exercise or a demonstration piece in any gathering.]

Note 3. Let p(b,n) be the probability that a family is of size

N=n and the number of brothers B=b, and suppose that the probability of selecting such a family is proportional to b. Then

$$p^w(b,n) = \frac{bp(b,n)}{E(B)} = \frac{bp(n)p(b|n)}{E(B)}, \qquad (5.11)$$

$$p^w(n) = \frac{E(B|n)}{E(B)}p(n). \qquad (5.12)$$

When $p(b|n)$ is binomial

$$p^w(n) = \frac{np(n)}{E(N)}, \quad E^w(1/N) = 1/E(N) \qquad (5.13)$$

so that the harmonic mean of observations $n_1, \ldots, n_k$ on N^w, i.e., from the distribution (5.11) or (5.12)

$$\frac{k}{\sum n_i^{-1}} \qquad (5.14)$$

is an estimate of E(N) in the original population. If the form of p(n) is known, then one could write down the likelihood of the sample n_1, ..., n_k using the probability function (5.12) and determine the unknown parameters by the method of maximum likelihood.

6. Alcoholism, family size and birth order

Smart (1963, 1964) and Sprott (1964) examined a number of hypotheses on the incidence of alcoholism in Canadian families using the data on family size and birth order of 242 alcoholics admitted to three alcoholism clinics in Ontario. The method of sampling is thus of the type discussed in Section 5.

One of the hypotheses tested was that *larger families contain larger number of alcoholics than expected*. The null hypothesis that the number of alcoholics is as expected was interpreted to imply that the observations on family size as ascertained arise from the weighted distribution

$$np(n)/E(N), n=1,2,...,$$
(6.1)

where $p(n)$, $n=1,2, ...$, is the distribution of family size in the general population. Smart and Sprott used the distribution of family size as reported in the 1931 census of Ontario for $p(n)$ in their analysis. It is then a simple matter to test whether the observed distribution of family size in their study is in accordance with the expected distribution (6.1).

It may be noted that the distribution (6.1) would be appropriate if we had chosen individuals (alcoholic or not) at random from the general population (of individuals) and ascertained the sizes of the families to which they belonged. But it is not clear whether the same distribution (6.1) holds if the inquiry is restricted to alcoholic individuals admitted to a clinic, as assumed by Smart and Sprott. This could happen, as demonstrated below, under an interpretation of their null hypothesis that the number of alcoholics in a family has a binomial distribution (like failures in a sequence of independent trials), and a further assumption that every alcoholic has the same independent chance of being admitted to a clinic.

Let π be the probability of an individual becoming an alcoholic, and suppose that the probability that a member of a family becomes an alcoholic is independent of whether another member is alcoholic or not. Further let $p(n)$, $n=1,2, ...$ be the probability distribution of family size (whether a family has an alcoholic of not) in the general population. Then the probability that a family is of size n and has r alcoholics is

$$p(n) \begin{bmatrix} n \\ r \end{bmatrix} \pi^r \phi^{n-r}, r=0,\ldots,n; \ n=1,2,\ldots, \qquad (6.2)$$

where $\phi=(1-\pi)$. From (6.2), it follows that the distribution of family size in the general population, given that a family has at least one alcoholic, is

$$\frac{(1-\phi^n)}{1-E(\phi^N)} p(n), n=1,2,\ldots \ . \qquad (6.3)$$

If we had chosen households at random and recorded the family sizes in households containing at least one alcoholic, then the null hypothesis on the excess of alcoholics in larger families could be tested by comparing the observed frequencies with the expected frequencies under the model (6.3). However, under the sampling scheme adopted of ascertaining the values of n and r from an alcoholic admitted to a clinic, the weighted distribution of (n,r),

$$p^w(n,r) = rp(n) \frac{n!}{r!(n-r)!} \frac{\pi^r \phi^{n-r}}{\pi E(N)} \qquad (6.4)$$

is more appropriate. If we had information on the family size n as well as on the number of alcoholics (r) in the family, we could have compared the observed joint frequencies of (n,r) with those expected under the model (6.4).

From (6.4), the marginal distribution of n alone is

$$np(n)/E(N), \ n=1,2, \ \ldots, \qquad (6.5)$$

which is used by Smart and Sprott as a model for the observed frequencies of family sizes. It is shown in (6.3) that in the general population, the distribution of family size in families with at least one alcoholic is

$$\frac{(1-\phi^n)p(n)}{1-E(\phi^n)},$$

which reduces to (6.5) if ϕ is close to unity. In other words, if the probability of an individual becoming an alcoholic is small, then the distribution of family size as ascertained is close to the distribution of family size in families with at least one alcoholic in the general population. This is not true if ϕ is not close to unity.

Smart and Sprott found that the distribution (6.5) did not fit the observed frequencies, which had heavier tails. They concluded that larger families contribute more than their expected share of alcoholics. Is this a valid conclusion? It is seen that the weighted distribution (6.5) is derived under two hypotheses. One is that the distribution of family size in the subset of families having at least one alcoholic in the general population is of the form (6.3) which is implied by the original null hypothesis posed by Smart. The other is that the method of ascertainment is equivalent to p.p.s. sampling of families, with probability proportional to the number of alcoholics in a family. The rejection of (6.5) would imply the rejection of the first of these two hypotheses if the second is assumed to be correct. There are no *a priori* grounds for such an assumption, and in the absence of an objective test for this, some caution is needed in accepting Smart's conclusions.

Another hypothesis considered by Smart was that the later born children have a greater tendency to become alcoholic than the earlier-born. The method used by Smart may be somewhat confusing to statisticians. Some comments were made by Sprott criticizing Smart's approach. We shall review Smart's analysis in the light of the model (6.4). If we assume that birth order has no relationship to becoming an alcoholic, and the probability of an alcoholic being referred to a clinic is independent of the birth order, then the probability that an observed alcoholic belongs to a family with n children and r alcoholics and has given birth order $s \le n$ is, using the model (6.4),

$$\frac{rp(n)}{n E(N)} \begin{pmatrix} n \\ r \end{pmatrix} \pi^{r-1} \phi^{n-r}, s=1,\dots,n; r=1,\dots,n; n=1,2,\dots. \quad (6.6)$$

Summing over r, we find that the marginal distribution of (n,s), the family size and birth order, applicable to observed data is

$$p(n)/E(N), \ s=1,\dots,n; n=1,2,\dots, \quad (6.7)$$

where it may be recalled that $p(n), n=1,2,\dots$, is the distribution of family size in the general population. Smart reported the observed bivariate frequencies of (n,s), and since p(n) was known, the expected values could have been computed and compared with the observed. But, he did something else.

From (6.7), the marginal distribution of birth ranks is

$$P(S=s) = \frac{1}{E(N)} \sum_{i=s}^{\infty} p(i), \ s=1,2,\dots. \quad (6.8)$$

Smart's (1963) analysis in his Table 2 is an attempt to compare the observed distribution of birth ranks with the expected under the model (6.8) with *p(i)* itself estimated from data using the model (6.1).

A better method is as follows: From (6.7) it is seen that for given family size, the expected birth order frequencies are equal as computed by Smart (1963) in Table 1, in which case individual chi-squares comparing the expected and observed frequencies for each family size would provide all the information about the hypothesis under test. Such a procedure would be independent of any knowledge of p(n). But it is not clear whether a hypothesis of the type posed by Smart can be tested on the basis of the available data without further information on the other alcoholics in the family, such as their age, sex, etc.

Table 4.6 reproduces a portion of Table 1 in Smart (1963)

relating to families up to size 4 and birth ranks up to 4. It is seen that for family sizes 2 and 3, the observed frequencies seem to contradict the hypothesis, and for family sizes above 3 (see Smart's Table 1), birth rank does not have any effect. It is interesting to compare the above with a similar type of date (Table 4.7) collected by the author on birth rank and family size of the staff members in two departments at the University of Pittsburgh. It appears that there are too many earlier-born among the staff members, indicating that becoming a professor is an affliction of the earlier born! It is expected that in data of the kind we are considering there will be an excess of the earlier born without implying an implicit relationship between birth order and a particular attribute, especially when it is age dependent. (This can be another classroom exercise. Go to any office and ascertain how

Table 4.6 Distribution of birth rank s and family size n
(Reproduced from Table 1 in Smart (1963))

s	$n=1$ O	E	2 O	E	3 O	E	4 O	E
1	21	21	22	16	17	13.3	11	11.75
2			10	16	14	13.3	10	11.75
3					9	13.3	13	11.75
4							13	11.75

O=observed, E=expected.

**Table 4.7 Distribution of birth ranks and family size $n \leq 4$
among staff members. (University of Pittsburgh)**

s	$n=1$	2	3	4
1	7	14	9	6
2		6	4	2
3			2	0
4				0

many are first born, second born, etc.. There will be a preponderance of the earlier born.)

7. Waiting time paradox

Patil (1984) reported a study conducted in 1966 by the Institute National de la Statistique et de l'Economie Appliquée in Morocco to estimate the mean sojourn time of tourists. Two types of surveys were conducted, one by contacting tourists residing in hotels and another by contacting tourists at frontier stations while leaving the country. The mean sojourn time as reported by 3,000 tourists in hotels was 17.8 days, and by 12,321 tourists at frontier stations was 9.0. Suspected by the officials in the department of planning, the estimate from the hotels was discarded.

It is clear that the observations collected from tourists while leaving the country correspond to the true distribution of sojourn time, so that the observed average 9.0 is a valid estimate of the mean sojourn time. It can be shown that in a steady state of flow of tourists, the sojourn time as reported by those contacted at the hotels has a size biased distribution, so that the observed average will be an over estimate of the mean sojourn time. If X^w is a size biased random variable (r.v.), then

$$E(X^w)^{-1} = \mu^{-1} \qquad (7.1)$$

where μ is the expected value of X, the original variable. The formula (7.1) shows that the harmonic mean of the size biased observations is a valid estimate of μ. Thus the harmonic mean of the observations from the tourists in hotels would have provided an estimate comparable with the arithmetic mean of the observations from the tourists at the frontier stations.

It is interesting to note that the estimate from hotel residents is nearly twice the other, a factor which occurs in the waiting time paradox (see Feller, 1966; Patil and Rao, 1977) associated with the exponential distribution. This suggests, but does not confirm, that sojourn time distribution may be exponential.

Suppose that the tourists at hotels were asked how long they had been staying in the country up to the time of inquiry. In such a case we may assume that the p.d.f. of the r.v. Y, the time a tourist has been in a country up to the time of inquiry, is the same as that of the product X^wR, where X^w is the size biased version of X, the sojourn time, and R is an independent r.v. with a uniform distribution on [0,1]. If F(x) is the distribution function of X, the p.d.f. of Y is

$$\mu^{-1}[1-F(y)].\qquad(7.2)$$

The parameter μ can be estimated on the basis of observations on Y, provided the functional form of F(y), the distribution of the sojourn time, is known.

It is interesting to note that the p.d.f. (7.2) is the same as that obtained by Cox (1962) in studying the distribution of failure times of a component used in different machines from observations on the ages of the components in use at the time of investigation.

8. Damage models

Let N be a r.v. with probability distribution, p_n, $n=1,2,\ldots$, and R be a r.v. such that

$$P(R=r\,|\,N=n)=s(r,n).\qquad(8.1)$$

Then the marginal distribution of R truncated at zero is

$$p_r' =(1-p)^{-1} \sum_{n=r}^{\infty} p_n s(r,n),\ r=1,2,\ldots,\qquad(8.2)$$

where

$$p = \sum_1^\infty p_i s(0, i). \tag{8.3}$$

The observation r represents the number surviving when the original observation n is subject to a destructive process which reduces n to r with probability s(r,n). Such a situation arises when we consider observations on family size counting only the surviving children (R). The problem is to determine the distribution of N, the original family size, knowing the distribution of R and assuming a suitable survival distribution.

Suppose that $N \sim P(\lambda)$, i.e., distributed as Poisson with parameter λ, and let $R \sim B(\pi)$, i.e., binomial with parameter π. Then

$$p'_r = e^{-\lambda\pi} \frac{(\lambda\pi)^r}{r! \ (1 - e^{-\lambda\pi})}, r = 1, 2, \ldots. \tag{8.4}$$

It is seen that the parameters λ and π get confounded, so that knowing the distribution of R we cannot find the distribution of N. Similar confounding occurs when N follows a binomial, negative binomial, or logarithmic series distribution. When the survival distribution is binomial, Sprott (1965) gives a general class of distributions which has this property. What additional information is needed to recover the original distribution? For instance, if we know which of the observations in the sample did not suffer damage, then it is possible to estimate the original distribution as well as the binomial parameter π.

It is interesting to note that the observations which do not suffer any damage have the distribution

$$p''_r = c p_r \pi^r, \tag{8.5}$$

which is a weighted distribution. If the original distribution is Poisson, then

$$p_r^u = e^{-\lambda \pi} \frac{(\lambda \pi)^r}{r ! (1 - e^{-\lambda \pi})} \, , \tag{8.6}$$

which is the same as (8.4). It is shown in Rao and Rubin (1964) that the equality $p_r^u = p_r'$ characterizes the Poisson distribution.

The damage models of the type described above were introduced in Rao (1965). For theoretical developments on damage models and characterization of probability distributions arising out of their study, the reader is referred to Alzaid, Rao and Shanbhag (1984).

References

Alzaid, A.H., Rao, C.R. and Shanbhag, D.N. (1984): Solutions of certain functional equations and related results on probability distributions. Technical Report, University of Sheffield, U.K.

Cox, D.R. (1962): *Renewal Theory*. Chapman and Hall, London.

Feller, W. (1966): *An Introduction to Probability Theory and its Applications*, Vol. 2, John Wiley & Sons, New York.

Feller, W. (1968): *An Introduction to Probability Theory and its Applications*, Vol. 1 (3rd edn.), John Wiley & Sons, New York.

Fisher, R.A. (1934): The effect of methods of ascertainment upon the estimation of frequencies. *Ann. Eugen.*, **6**, 13-25.

Patil, G.P. (1984): Studies in statistical ecology involving weighted distributions. In *Statistics: Applications and New Directions*, 478-503. Indian Statistical Institute, Calcutta.

Patil, G.P. and Ord, J.K. (1976): On size-biased sampling and related form-invariant weighted distributions. *Sankhyā* Ser. B **33**, 49-61.

Patil, G.P. and Rao, C.R. (1977): The weighted distributions: A survey of their applications. In *Applications of Statistics* (P.R. Krishnaiah, Ed.), 383-405, North Holland Publishing Company, Amsterdam.

Patil, G.P. and Rao, C.R. (1978): Weighted distributions and size biased sampling with applications to wildlife populations and human families. *Biometrics*, **34**, 170-180.

Rao, C.R. (1965): On discrete distributions arising out of methods of ascertainment. In *Classical and Contagious Discrete Distributions*, (G.P.

Patil, Ed.), 320-33. Statist. Publishing Society, Calcutta. Reprinted in *Sankhyā* Ser. A, **27**, 311-324.

Rao, C.R. (1973): *Linear Statistical Inference and its Applications,* (2nd Edn.), John Wiley & Sons, New York.

Rao, C.R. (1975): Some problems of sample surveys. *Suppl. Adv. Appl. Probab.,* **7**, 50-61.

Rao, C.R. (1977): A natural example of a weighted binomial distribution. *Amer. Statist.,* **31**, 24-26.

Rao, C.R. (1985): Weighted distributions arising out of methods of ascertainment: What population does a sample represent? In a *Celebration of Statistics*, the ISI Centenary Volume (A.C. Atkinson and S.E. Fienberg, Eds.), 543-569. Springer-Verlag.

Smart, R.G. (1963): Alcoholism, birth order, and family size. *J. Abnorm. Soc., Psychol.,* **66**, 17-23.

Smart, R.G. (1964): A response to Sprott's "Use of Chi-square". *J. Abnorm. Soc. Psychol.,* **69**, 103-105.

Sprott, D.A. (1964): Use of Chi-square. *J. Abnorm. Soc. Psychol.,* **69**, 101-103.

Sprott, D.A. (1965): Some comments on the question of identifiability of parameters raised by Rao. In *Classical and Contagious Discrete Distributions* (G.P. Patil, Ed.), 333-336. Statist. Publishing Society, Calcutta.

Chapter 5

Statistics: an Inevitable Instrument in Search of Truth

1. Statistics and truth

> *But as for certain truth, no man known it,*
> *Nor will he know it; neither of the gods,*
> *Nor yet of the things of which I speak.*
> *And even if he were by chance utter*
> *The final truth, he would himself not know it;*
> *For all is but a woven web of guesses.*
> Xenophanes of Kolophon

In the first two chapters, I referred to uncertainty in our real world. Uncertainty may arise through lack of information, lack of sufficient knowledge in utilizing available information, errors in measurements even using sophisticated instruments, acts of God (catastrophes), vagaries of human behaviour (the most unpredictable of all phenomena), random behaviour of fundamental particles requiring probabilistic rather than deterministic laws in explaining natural phenomena, etc. I mentioned how quantification of uncertainty enables us to devise methods to reduce, control or take uncertainty into account in making decisions. In third and fourth chapters I discussed strategies of data analysis for extracting information from observed data and dealing with uncertainty. I emphasized the need to have clean, relevant and honest data and to use appropriate models in extracting information. In this chapter I shall pursue the same theme a little further and discuss through some examples the role of statistics in the wider context of acquiring new knowledge or searching for truth to understand nature and taking optimal decisions in our daily life.

What is knowledge and how do we acquire it? What are the thought processes involved and the nature of investigations to be carried out? These questions have baffled the human intellect and remained for a long time the subject of philosophical discourses. However, recent advances in logic and statistical science opened up a systematic way of acquiring new knowledge, interpreted in a pragmatic rather than the metaphysical sense of "true knowledge."

1.1   *Scientific Laws*

Scientific laws are not advanced by the principal of authority or justified by faith or medieval philosophy; statistics is the only court of appeal to new knowledge.

P.C. Mahalanobis

A beautiful theory, killed by a nasty, ugly little fact.

Thomas H. Huxley

Science deals with knowledge of natural phenomena and its improvement. Such knowledge is usually abstracted in terms of laws (axioms or theories) which enable prediction of future events within requisite limits of accuracy and which provide the basis for technological research and applications. Thus, we have Newton's laws of motion, Einstein's theory of relativity, Bohr's atomic model, Raman effect, Mendel's laws of inheritance, double helix DNA, Darwin's theory of evolution, etc., on which the modern technology depends. We may never know what the true laws are. Our search is only for working hypotheses which are supported by observational facts and which, in course of time, may be replaced by better working hypotheses with more supporting evidence from a wider set of data and provide wider applicability. We study the world as it seems to be. "It does not matter to science whether there are really electrons or not provided things behave as if there were" (Macmurray, 1939). The scientific method of investigation involves the following endless cycle (or spiral) which is an elaboration of Popper's formula ($P_1 \rightarrow TT \rightarrow EE \rightarrow P_2$) where P_1, P_2 stand for initial theory and its modification respectively, TT for testing theory and EE for elimination of errors.

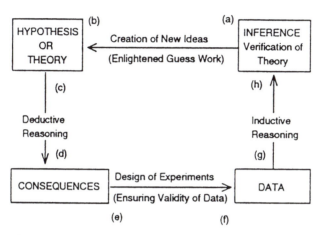

Every hypothesis is possibly rejected with the accumulation of more data, a situation bluntly described by Karl Popper:

> Supporting evidence for a scientific hypothesis is merely an attempt at falsification which failed.

The scientific method as shown in the above diagram involved two logical processes - deductive reasoning and inductive reasoning. A detailed discussion of the difference between two is given in Chapter 2.

As shown in the above diagram, there are two phases in the scientific method: paths (a)→(b) and (c)→(d) come under the subject field of research and the creative role played by the scientist and the other paths (e)→(f) and (g)→(h) come under the realm of statistics. Through collection of relevant and valid data by efficiently designed experiments and appropriate data analysis to test given hypotheses and to provide clues for possible alternatives, statistics enables the scientist to have a full play for his creative imagination to discover new phenomena without allowing them to run riot and waste in advancing new concepts which have no relation to existing facts. Statistical methods have been of great value especially in biological and social sciences where the range of variation in observations is often large and the number of observations is often limited; only statistical analysis can give a quantitative estimate of the significance of the findings in such situations.

Commenting on the importance of designing an efficient experiment in scientific work (path (e)→(f) in the above diagram), using statistical principles, R.A. Fisher (1957) says,

> A complete overhauling of the process of the collection, or of experimental design, may often increase the yield ten or twelve fold, for the same cost in time and labor. To consult a statistician after an experiment is finished is often merely to ask him to conduct a *post mortem* examination. He can perhaps say what the experiment died of.

1.2   *Decision Making*

> *To guess is cheap, to guess wrongly is expensive.*
>   An old Chinese proverb

In decision making we have to deal with uncertainty. The nature of uncertainty depends on the problem. Typical questions leading to decision making are as follows. How much corn will be produced in the current year? Is the accused person in a certain case guilty? Is a woman's claim that a particular person is the father of her child correct? Does smoking cause lung cancer? Does one aspirin tablet taken every other day reduce the risk of heart attack? Was a particular skull, found in an ancient grave, that of a man or of a woman? Who wrote the play, *Hamlet*, Shakespeare, Bacon or Marlowe? What is the exact location of the brain tumor in a patient's head? What is the family tree of the different languages in the world?

Is the last born child more or less intelligent than the first born? What will be the price of gold two months from now? Does the use of a seat belt protect the driver of an automobile from serious injuries in an accident? Do the planets control our movements, actions and achievements? Are astrological predictions correct?

These are all situations which cannot be resolved by philosophical discussions or by using existing (or established) theories. No definite answers can be found from available information or data, and any prescribed rule for selecting one out of possible answers will be subject to error. *The alternative to avoiding mistakes is not refraining from taking decisions.* There can be no progress that way. The best we can do is to take decisions in an optimal way by minimizing the risk involved. We discuss a number of examples where statistics enabled to resolve the issues involved.

1.3   *The Ubiquity of Statistics*

> *Statistical science is the peculiar aspect of human progress which gave the 20th century its special character, ... it is to the statistician that the present age turns for what is most essential in all its more important activities.*
>
> R.A. Fisher (1952)

The scope of statistics as it is understood, studied and practiced today extends to the whole gamut of natural and social sciences, engineering and technology, management and economic affairs and art and literature. The ubiquity of statistics is illustrated in the following chart.

The *layman* uses statistics (information obtained through data of various kinds and their analyses published in newspapers and consumer reports) for taking decisions in daily life, or making future plans, deciding on wise investments in buying stocks and shares, etc. Some amount of statistical knowledge may be necessary for a proper understanding and utilization of all the available information and to

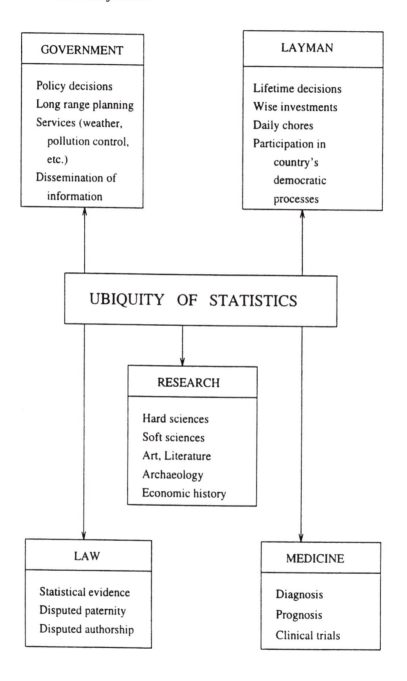

guard oneself against misleading advertisements. The need for statistical literacy in our modern age dominated by science and technology was foreseen by H.G. Wells:

> Statistical thinking will one day be as necessary for efficient citizenship as the ability to read and write.

For the *government* of a country, statistics is the means by which it can make short and long range plans to achieve specified economic and social goals. Sophisticated statistical techniques are applied to make forecasts of population and the demand for consumer goods and services and to formulate economic plans using appropriate models to achieve a desired rate of progress in social welfare. It is said, "The more prosperous a country is, the better is its statistics." This, indeed, is a statement where the cause and effect are reversed. With vast amounts of socio-economic and demographic data now collected through administrative channels and special sample surveys and advances in statistical methodology, public policy making is no longer a gamble with an unpredictable chance of success or a hit and miss affair. It is now within the realms of scientific techniques whereby optimal decisions can be taken on the basis of available evidence and the results continuously monitored for feedback and control.

In *scientific research*, as I have mentioned earlier, statistics plays an important role in the collection of data through efficiently designed experiments, in testing hypotheses and estimation of unknown parameters, and in interpretation of results. The discovery of the Rhesus factor in blood groups, as described by Fisher (1947), is a brilliant example of how statistics can be of help in fitting one scrupulously ascertained fact into the other, in building a coherent structure of new knowledge and seeing how each gain can be used as a means for further research (see subsection 2.18 in this chapter).

In *industry*, extremely simple statistical techniques are used to improve and maintain the quality of manufactured goods at a desired level. Experiments are conducted in R. & D. departments to determine the optimum mix (combinations of factors) to increase the yield or give the best possible performance. It is a common experience all over the world that in plants where statistical methods are exploited, production has increased by 10% to 100% without further investment or expansion of plant. In this sense statistical knowledge is considered as a national resource. It is not surprising that a recent book on modern inventions lists statistical quality control as one of the great technological inventions of the present century.

Indeed, there has rarely been a technological invention like *statistical quality control*, which is so wide in its application yet so simple in theory, which is so effective in its results yet so easy to adopt and which yields so high a return yet needs so small an investment.

In *business*, statistical methods are employed to forecast future demand for goods, to plan for production, and to evolve efficient management techniques to maximize profit.

In *medicine*, principles of design of experiments are used in screening of drugs and in clinical trials. The information supplied by a large number of biochemical and other tests is statistically assessed for diagnosis and prognosis of disease. The application of statistical techniques has made medical diagnosis more objective by combining the collective wisdom of the best possible experts with the knowledge on distinctions between diseases indicated by tests.

In *literature*, statistical methods are used in quantifying an author's style, which is useful in settling cases of disputed authorship.

In *archeology*, quantitative assessment of similarity between objects has provided a method of placing ancient artifacts in a chronological order.

In *courts of law*, statistical evidence in the form of probability of occurrence of certain events is used to supplement the traditional oral and circumstantial evidence in judging cases.

In *detective work*, statistics helps in analyzing bits and pieces of information, which individually may appear to be unrelated or even inconsistent, to see an underlying pattern. An interesting case study of this nature can be found in the book *A Perfect Spy* by John Le Carre, where data on "names of all their contacts, details of their travel movements, behaviour of their contacts, sexual and recreational appetites" enables certain conclusion to be drawn on the spying activities of some individual by relating these data to certain events.

There seems to be no human activity whose value cannot be enhanced by injecting statistical ideas in planning and by using statistical methods for efficient analysis of data and assessment of results for feedback and control. It is apodictic to claim: If there is a problem to be solved, seek for statistical advise instead of appointing a committee of experts. Statistics and statistical analysis can throw more light than the collective wisdom of the articulate few.

2.   Some examples

I shall give you a number of examples drawn from the story of "the improvement of natural knowledge" and the success of "decision making" to show how statistical ideas played an important role in scientific and other investigations even before statistics was recognized as a separate discipline and how statistics is now evolving as a versatile, powerful and indispensable instrument for investigation in all fields of human endeavor.

2.1   *Shakespeare's poem: An ode to statistics*

Not marble, nor gilded monument of princes, shall out live this powerful rhyme.

Shakespeare

On 14 November 1985, the Shakespearean scholar Gary Taylor found a nine-stanza poem in a bound folio volume that has been in the collection of the Bodelian Library since 1775. The poem has only 429 words and there is no record as to who was the author of the poem. Could it be attributed to Shakespeare? Two statisticians, Thisted and Efron (1987) made a statistical study of the problem and concluded that the poem fits nicely with Shakespeare's style (canon) in the usage of words. The investigation was based on a purely statistical study as follows.

The total number of words in all the known works of Shakespeare is 884,647 of which 31,534 are distinct and the frequencies with which these words were used are given in Table 5.1. The information contained in the Table 5.1 can be used to answer questions of the following kind. If Shakespeare were asked to write a new piece of work consisting of a certain number of words, how many new words (not used in earlier works) would he use? How many words will there be, which he used only once, twice, thrice, ...

Table 5.1: Frequency distribution of usage of distinct words

No. of times a word is used	No. of distinct words
1	14,376
2	4,343
3	2,292
4	1,463
5	1,043
6	837
7	638
.	.
.	.
.	.
> 100	846
TOTAL	31,534

in all his earlier works? It is possible to predict these numbers using a remarkable law described by R.A. Fisher *et al* (1943), in an entirely different area, for estimating the total number of unseen species of butterflies! Using Fisher's theory, it was estimated that Shakespeare would have used about 35,000 new words if he were to write new dramas and poems containing the same number of words 884,647 as in his previous works. This would place the total vocabulary of Shakespeare at an estimated level of more than 66,000 words. [At the time of Shakespeare, there were about 100,000 words in the English language. At present there are about 500,000 words.]

Now coming back to the newly discovered poem, which has 429 words of which 258 are distinct, the observed and predicted (according to Shakespearean canon) distributions are as given in Table 5.2 (last two columns). It is seen that the agreement between the two distributions is quite close (within the limits of expected difference) suggesting that Shakespeare was the possible author of the poem.

Table 5.2 also gives similar frequency distributions of words in poems of about the same size by other contemporary authors, Ben Johnson, Christopher Marlowe and John Donne. The frequencies in the case of these authors look somewhat different from observed frequencies in the new poem and also the predicted frequencies under Shakespearean usage of words.

2.2 *Disputed authorship: The Federalist papers*

A closely related problem is that of disputed authorship or the identification of the author of an anonymous work from a possible panel of authors, I shall give you an example of such an application. The method employed is due to Fisher, who first developed it in an answer to a question put to him by an anthropologist. Is there an objective way, using measurements alone, of deciding whether a mandible recovered from a grave was that of a man or of a woman?

The same technique can be used to answer essentially a similar question: Which one of two possible writers authored a

Table 5.2 Frequency distributions of distinct words in poems according to Shakespearean canon in poems of similar length by different authors

Number of times used in Shakespeares works	Number of distinct words used in				Expected according to Shakesperian cannon
	Ben Johnson (An Elogy)	Christopher Marlov (four Poems)	John Donne (The Ecstacy)	New poem	
0	8	10	17	9	6.97
1	2	8	5	7	4.21
2	1	8	6	5	3.33
3-4	6	16	5	8	5.36
5-9	9	22	12	11	10.24
10-19	9	20	17	10	13.96
20-29	12	13	14	21	10.77
30-39	12	9	6	16	8.87
40-59	13	14	12	18	13.77
60-79	10	9	3	8	9.99
80-99	13	13	10	5	7.48
No. of distinct words ..	243	272	252	258	258
Total No. of words ..	411	495	487	429	-

disputed piece of work? Let us consider the case of the *Federalist Papers* written during the period 1787-1788 by Alexander Hamilton, John Jay and James Madison to persuade the citizens of New York to ratify the constitution. There were 77 papers signed with a pseudonym "Publicus" as was common those days. The exact authorship of many of these essays have been identified, but the authorship of 12 was in dispute between Hamilton and Madison. Two statisticians, Frederic Mosteller and David Wallace (1964) came to the conclusion, using a statistical approach, that Madison was the most likely author of the 12 disputed papers. The quantitative approach in such cases is to study each individual author's style from his known publications and to assign the disputed work to that author whose style is closest to the disputed work.

2.3   *Kautilya and the Arthaśāstra*

The *Kautilya Arthaśāstra* is regarded as a unique work, which throws more light on the cultural environment and the actual life in ancient India than any other work of Indian literature. This remarkable treatise is considered to be written in the fourth century B.C. by Kautilya, the minister of the famous king Chandragupta Maurya. However, various scholars have raised doubts both about the authorship of the text *Arthaśāstra* as well as the period of its publication.

Some years ago Trautman (1971) made a statistical investigation of the authorship and date of publication of *Arthaśāstra*. He found considerable variation in the styles of prose in different parts of the book and came to the conclusion that Kautilya could not have been the sole author of *Arthaśāstra* and that it must have been written by several authors, perhaps three or four, at different periods of time, centering around the middle of second century A.D. Since there are no other known works of Kautilya, it is difficult to say which part he wrote, if at all he made any contribution to it.

2.4   *Dating of publications*

When did Shakespeare write *Comedy of Errors* and *Love's Labors Lost*? The dates of publication of most of Shakespeare's works are known through written records, but in some cases they are not. How can the information about the known dates of some publications be used to estimate the unknown dates of other publications? Yardi (1946) examined this problem by using a purely quantitative method and no external evidence. For each play, he obtained the frequencies of: (i) redundant final syllables, (ii) fullsplit lines, (iii) unsplit lines with pauses and (iv) the total number of speech lines. With the literary style so quantified, Yardi studied the secular changes in style over the long period of Shakespeare's literary output using the data on plays with known dates of publication. He then inferred by interpolation the possible date of publication of *Comedy of Errors* as in the winter of 1591-92 and that of *Love's Labors Lost* as in the spring of 1591-92.

2.5   *Seriation of Plato's works*

Plato's works survived for more than 22 centuries and his philosophical ideas and elegant style have been widely studied. Unfortunately, nobody mentioned or perhaps nobody knew the correct chronological order in which his 35 dialogues, 6 short pieces and 13 letters appeared. The problem of chronological seriation of Plato's works was posed a century ago but no progress was made. The statisticians took up the problem a few years ago and have now provided what appears to be a logical solution.

The statistical method starts by establishing for each pair of works an index of similarity. In a study undertaken by Boneva (1971), the index was based on the frequency distribution in each work, of 32 possible descriptions of the last 5 syllables of a sentence, technically called *Clausula*. Using the only assumption that works closer in time would be similar in style, and no other extraneous information, a method has been evolved to infer the chronological order of Plato's works.

2.6   *Filiation of manuscripts*

Filiation or linkage of manuscripts is another problem solved by purely statistical techniques. A recent study by Sorin Christian Nita (1971) related to 48 copies of the Romanian chronicle, *The

History of Romania, some of which are copied from the original, and the others from the copies of one or more removed from the original. The problem was to decide, as far as possible, the original version of the work and the whole genealogical tree of the existing manuscripts. Here, the statistician exploits the human failing of making errors while copying from a given manuscript. Thus, although the manuscripts are all of the same original work, they differ in errors and possible alterations made while copying. An error in a manuscript is propagated to all its descendants and two copies made from the same manuscript have more common errors than those copied from different manuscripts. Using the number of common errors between each pair of manuscripts as the only basic data, it has been possible to work out the entire linkage of the manuscripts.

2.7   *The Language tree*

By studying the similarities between the Indo-European languages (consisting of such diverse languages as of Latin and Sanskrit origin, Germanic, Slavic, Baltic, Iranian, Celtic, etc.), linguists have discovered a common ancestor which is believed to have been spoken about 4,500 years ago. And if there is a common ancestor, there must also be an evolutionary tree of the languages branching off at different points of time. Is it possible to construct such a language tree similar to the evolutionary tree of life constructed by the biologists? This is, indeed, an exciting and challenging problem, and the scientific study of such problems is called "Glotto chronology." Using a vast amount of information on similarities between languages and a complicated reasoning, linguists were able to identify some major branches of languages, but the exact relationships between them and the times of separation could not be well established. However, a purely statistical approach to this problem using less information has given very encouraging results.

A first step in such a study is the comparison of words belonging to different languages for a basic set of meanings such as eye, hand, mother, one, ... and so on. Words with the same meaning belonging to two different languages are scored with + sign if they cognate and − sign otherwise. Thus a comparison of two languages is expressed as a sequence of + and − signs or a vector of the form $(+, −, +, +, ...)$. If there are n languages, there are $n(n-1)/2$ such similarities. Using this information only, Swadish (1952) suggested a method of estimating the time of separation between two languages. Once the times of separation of all pairs of languages are known, it is easy to construct an evolutionary tree. The whole task is simplified and made routine by suitable computer programs designed to print out the whole evolutionary tree by feeding in the comparison vectors of + and − signs. The method was recently applied to construct evolutionary trees of the Indo-European languages using a list of 200 meanings and Malayo-Polynesian languages using a list of 196 meanings (Kruskal, Dyen and Black, (1971)).

In the application of statistics to literature, such as dating of Shakespeare's works, chronology of Plato's works, linkage of manuscripts, etc., one may question the validity of the results (or the method employed). The logical issues involved are the same as when you ask the question: how good are Paraxin tablets for a particular patient for curing his typhoid fever? The only justification is that these tablets helped many typhoid patients before. But could they not be fatal to a particular patient? In the same way, the validity of a statistical method is established by what is called a "performance test." A proposed method is first used to predict in some known cases and the method is accepted only when its performance is found to be satisfactory. Of course, one should always look for independent historical and other evidences, if available, to corroborate the statistical findings.

2.8   *Geological time scale*

This is an example quoted by Fisher (1952) to illustrate the statistical thinking behind one of the greatest discoveries in geology.

We are all familiar with the geological time scale and the names of geological strata like Pliocene, Miocene, Oligocene etc., but many may not be aware how these were arrived at. This was the brain child of the geologist Charles Lyell who was born in 1797 and wrote the celebrated book *Principles of Geology*. In the third volume of this book issued in 1833, he gave detailed calculations of these time scales, which represent a highly sophisticated statistical approach based on a completely novel idea.

With the aid of the eminent conchologist M. Deshayes, Lyell proceeded to list the identified fossils occurring in one or more strata, and to ascertain the proportions now living. It was as though a statistician has a recent census record without recorded ages and a series of undated records of previous censuses in which some of the same individuals could be recognized. A knowledge of the Life Table would then give him estimates of the dates, and even without a Life Table, he could set the series in a chronological order, merely by comparing the proportion in each record of those who were still living; the older the formation of a strata, the smaller will be the proportion of fossils still living. Lyell's thinking and the superb statistical argument by which he named different strata and which brought about a revolution in Geological Science is illustrated in the Table 5.3. With the aid of such a classification, geologists could recognize a fossiliferous stratum by a few characteristic forms with clear morphological peculiarities. Unfortunately, the quantitative thinking behind Lyell's method is never emphasized in courses given to students.

2.9   *Common breeding-ground of eels*

This is an example taken from Fisher (1952) to illustrate how elementary descriptive statistics led to an important discovery.

In the early years of the present century, Johannes Schmidt of the Carlsberg Laboratory in Copenhagen found that the numbers of vertebrae and fin rays of the same species of fish caught in different

Table 5.3 Lyell's geological classification

Name given to geological strata	Percentage of surviving species	Examples
PLEISTOCENE (most recent)	96%	Sicilian Group
PLIOCENE (majority recent)	40%	Sub-appenine Italian Rocks, English Crag
MIOCENE (minority recent)	18%	"
EOCENE (dawn of the recent)	3% or 4%	"
"	"	"

localities varied considerably; often even from different parts of the same fjord. With eels, however, in which the variation in vertebrate number is large, Schmidt found sensibly the same mean, and the same standard deviation, in samples drawn from all over Europe, from Iceland, from the Azores and from the Nile, which are widely separated regions. He inferred that the eels of all these different river systems came from a common breeding-ground in the ocean, which was later discovered in one of the expeditions of the research vessel "Dana".

2.10   *Are acquired characteristics inherited?*

This question arose in a discussion on Darwin's theory and, in order to answer this, a Danish geneticist W. Johannsen conducted an experiment which might appear as a textbook stuff now-a-days but not in 1909 when Johannsen first published his results. I quote from a note by Marc Kac (1983) who was introduced to this subject when he was 13 years old.

"Johannsen took a large number of beans, weighed them, and on the basis of these weights constructed a histogram to which he fitted a normal curve which is now well known. Having done this, he took the smaller beans and the large ones, planted them separately, and constructed histograms of the weights of their respective progeny. These he again fitted with normal curves. If size were inheritable, one could expect the two curves to be centered on different means - the small and the large. As it turned out, the two curves were essentially indistinguishable from the original parental curve, thus raising serious doubts as to the inheritability of smallness or largeness." Kac continues:

"What struck me at that time, and remains with me today, was the utter novelty of the argument, which was unlike anything I have encountered up to that time in mathematics, physics, or biology. I have since learned a good deal of statistics and have even taught it at levels requiring varying degrees of mathematical sophistication, but I still consider Johannsen's experiment one of the best illustrations I know of the power and elegance of statistical reasoning."

2.11   *The importance of being left-handed*

It is not generally known that a coconut tree can be classified as left-handed or right-handed, depending on the direction of its foliar spiral. Some years ago, an investigation of this aspect was undertaken by T.A. Davis at the Indian Statistical Institute (ISI). The study offers a good example of a statistical approach in understanding nature, where observational facts suggest new problems, in solving which further observations are made. The gains attained at each stage are consolidated and fresh evidence is sought to strengthen the basis of earlier results and to explore new aspects.

Why are some trees left-handed and others right-handed? Is

this character genetically inherited? The question can be answered by considering parent plants of different combinations of foliar spirality and scoring the progeny for the same characteristic. The data collected for this purpose are shown in Table 5.4. The ratios of left to right are nearly the same for all combinations of parents indicating that there is no genetic basis for left- or right-handedness.

So the ratio appears to be entirely determined by external factors which act in a random way. But why is there a slight preponderance of right-handed offsprings (about 55 percent) in the observed data (Table 5.4)? There must be something in the environment which tends to give a greater chance for a tree to twist in the right direction. And if so, does this chance depend on the

Table 5.4 Proportions of left- and right-handed offsprings for different types of mating

Pollen parent	Seed parent	Progeny		
		left	:	right
Right	Right	44	:	56
Right	Left	47	:	53
Left	Right	45	:	55
Left	Left	47	:	53

geographical location of trees? This could not be determined until data from various parts of the world could be collected. It was then found that the proportion of left-handers is 0.515 in samples from the Northern Hemisphere, and 0.473 in the Southern Hemisphere. The difference may be due to the influence of the one-way rotation of the Earth, which also explains the phenomenon of the bathtub vortex (the left or right spiral in which water drains out of a bathtub when the stopcock is removed) which, under well-controlled conditions, is shown to be more frequently counter-clockwise in the Northern Hemisphere and more frequently clockwise in the Southern Hemisphere.

The investigations would have remained somewhat academic in character if Davis had not been curious to look for some features in which the left and right trees could possible differ. He compared the mean yields of left and right trees in a plantation over a 12-year period; he was surprised to find that the former yielded 10 percent more than the latter. Although no explanation could be offered - the question needs to be pursued and might not be easily solved - the empirical conclusion is of great economic importance. For by selective planting of left trees alone, the yield could be increased by 10 percent! Davis has raised the question whether left-handed woman are more fertile than the right-handed. A study by the Sanford Corporation suggests that left-handers tend to be exceptionally creative and good looking. It says that there are all sorts of lefties of whom the left-handers can be proud: Benjamin Franklin, Leonardo da Vinci, Albert Einstein, Alexander the Great, Julius Ceaser, ...

The phenomenon of right- and left-handedness seems to be universal in the plant kingdom. You may not have noticed flowers with right and left spiral arrangement of petals on the same plant in your garden. And there are creepers which twine up only in a right spiral and others only in a left spiral. Experiments at the Indian Statistical Institute, Calcutta, to change their habits ended in a failure. They seem to react violently at any such attempt.

It is also strange that all living organisms (expect possibly very low forms) are left-handed in their biochemical make-up. All amino-acids except glycine, exist in two forms - L (levo) and D (dextro). The L and D forms are mirror images of each other and are called the left- and right-handed molecules, respectively. All the 24 amino-acids found in plant and animal proteins and even in simple organisms like bacteria, moulds, viruses, etc. are left-handed. Both right-handed and left-handed molecules have exactly the same properties, and life might have been possible with only D acids or even with a mixture of some L and some D acids. Is it then an accident of nature that living organisms have evolved in the L-system rather than in the D-system? Or, is it possible that the left-handed molecules are inherently more suited to the construction of living organisms? There may be some mysterious force in leaning to left, which science has yet to explore.

The diagram reproduced below with the permission of the late Dr. T.A. Davis of the Indian Statistical Institute illustrates the left and right spirality of stems in plants and petals in flowers.

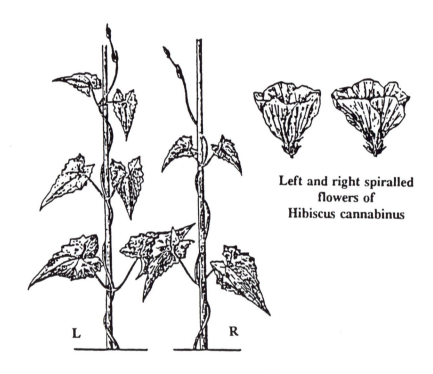

Left and right spiralled
flowers of
Hibiscus cannabinus

Left and right spiralling stems of Mikamia Scanders

Dr.Robert Sperry, the Nobel prize winner, established that in each individual either the left or the right brain dominates, the left brain people being more in number. It appears that the simplest way to characterize a right-brain person would be by his creative ability, whereas the left-brain person would be more logical.

2.12   *Circadian rhythm*

If you are asked, what is your height, you will, no doubt, have a ready answer - a certain number. Someone might have measured your height some time and given you that number. But you might not have enquired how that number qualified to represent your height. And if you, indeed, had, the answer would have been, that it is an observation obtained by carefully following a "prescribed procedure for measuring height." For all practical purposes such an operational definition of height may be satisfactory. But then other questions arise. Does the characteristic we are trying to measure (in a prescribed way) depend on the time of the day at which the measurement is taken? And, if it is variable, how do we specify its value? For instance, is there a difference between morning and evening (true) heights of an individual? If so, what is the magnitude of the difference and does it have any physiological explanation?

A simple statistical investigation can provide the answer. Careful determinations of morning and evening heights of 41 students in Calcutta showed an average difference of 9.6 mm, the morning measurement being higher in each case (see Rao, (1957)). If, in fact, the height of an individual at different times of the day is the same, then any observed difference is attributable to errors of measurement which may be positive or negative with equal probability. In such a case, the probability that all the 41 differences are positive is of the order of 2^{-41}, which corresponds to an event which occurs less than 5 times in 10^{13} experiments, indicating that the odds against the hypothesis of no difference in heights are extremely high. We seem to grow by about 1 cm when we are asleep at night and diminish by the same length while we are at work during the day!

Having established that the morning and evening heights

differ, the next question may be, which part of the body elongates more when we are asleep? To examine this, separate determinations were made of the lengths between certain points marked on the body, both in the morning and in the evening. It was found that the entire difference of about 1 cm occurred in that part of the body along which the vertebral column is located. A plausible physiological explanation is that during the day the vertebrae come closer by shrinkage of the cartilages between them; they revert to the original position when the body is relaxed.

Why do teachers prefer to lecture in the morning hours? It is said that both teachers and students are fresh in the morning and there is greater rapport between them. Is there any physiological explanation of this phenomenon?

The change in the plasma levels seems to explain our alertness in the morning hours. In normal subjects, the cortisol level is about 16 mg/100 ml at 8 a.m. and it gradually drops to 6 mg/100 ml by 11 p.m. (a decrease of 60 percent). The rise of cortisol in the morning wakes you up and the trough in the evening puts you to sleep. Consequently we are alert in the morning and gradually tend to be sluggish as the night falls.

Several physiological characteristics of the human body, in fact, vary during the day as was observed in the case of the height; each has a particular circadian rhythm, that is, it follows a 24-hour cycle. The importance of studying such variations, known as Chronobiology, for optimum timing of administering medicines to patients has been stressed by Halberg (1974). For instance, a dose of a drug which is right at one time of the day can be found to be not effective at another time; the action may depend on the levels of different biochemical substances in the blood at the time of administering the drug. Chronobiology is becoming an active field of research with extensive possibilities of application. Much progress in these studies is due to statistical techniques developed to detect and establish periodicities in measurements taken over time.

2.13 *Disputed paternity*

Suppose, a mother says that a certain man is the father of her child, and the man denies it. Can we compute the chance of the accused man being the father, which could be used in a court of law, possibly along with other evidence to decide the case? In many countries, courts of law accept statistical evidence in deciding cases of disputed paternity.

Usually, the evidence is based on matching the blood groups or DNA sequences. In certain cases the blood groups or DNA sequences of the putative father and child may not be compatible leading to a definite conclusion that the mother's claim is wrong. However, if the blood groups or DNA sequences are compatible, this does not imply that the claim is correct. In such a case, we can compute the probability of the claim being correct. If this is high then there may be a case for accepting the claim if there is support from other evidence.

2.14 *Salt in statistics*

> *...and, what is still extraordinary, I have met with a philosophical work in which the utility of salt has been made the theme of an eloquent discourse, and many other like things have had a similar honour bestowed on them.*
> Pheadrus (Plato's Symposium on Love)

There were communal riots in Delhi in 1947 immediately after India achieved independence. A large number of people of a minority community took refuge in the Red Fort which is a protected area, and a small number in the Humanyun Tomb, another area enclosing an ancient monument. The Government had the responsibility to feed these refugees. This task was entrusted to contractors, and in the absence of any knowledge about the number of refugees, the government was forced to accept and pay the amounts quoted by the contractors for different commodities purchased by them to feed the refugees. The government expenditure on this account seemed to be extremely high and it was suggested that statisticians (who count) may be asked to find the number of refugees inside the Red Fort.

The problem appeared to be difficult under the troubled conditions prevailing at that time. A further complication arose as the statistical experts called in to do the job belonged to the majority community (different from that of the refugees) and their safety could not be guaranteed if the statistical techniques to be applied by them for estimating the number of refugees required their getting inside the Red Fort. Then the problem before the experts was to estimate the number of persons inside a given area without any prior information about the order of magnitude of the number, without having any opportunity to look at the concentrations of persons inside the area and without using any known sampling techniques for estimation or census methods.

The experts had to think of some way of solving the problem. Giving up might be interpreted by the government as failure of statistics and/or of the statisticians. They had, however, access to the bills submitted by the contractors to the government, which gave the quantities of various commodities such as rice, pulses and salt purchased by them to feed the refugees. They argued as follows?

Let R, P and S represent the quantities of rice, pulses and salt used per day to feed all the refugees. From consumption surveys, the per capita requirements of these commodities are known, say, r, p and s respectively. Then R/r, P/p and S/s would provide parallel (equally valid) estimates of the *same number* of persons. When these ratios were computed using the values R, P and S quoted by the contractors it was found that S/s had the smallest value and R/r, the largest value indicating that the quantity of rice, which is the most expensive commodity compared to salt, was probably exaggerated. (The price of salt was extremely low in India at that time and it would not pay to exaggerate the amount of salt). The estimate S/s was proposed by the statisticians for the number of refugees in the Red

Fort. The proposed method was verified to provide a good approximation to the number of refugees in the Humayun tomb (the smaller of the two camps with only a relatively small number of refugees), which was independently ascertained.

The salt method arose out of an idea suggested by the late J.M. Sengupta who was associated with the Indian Statistical Institute for a long time. The estimate provided ny the statisticians was useful to the government in taking administrative decisions. It also enhanced the prestige of statistics which received good government support ever since, for its development in India.

The method used is unconventional and ingenious, not to be found in any textbook. The idea behind it is statistical reasoning or quantitative thinking. Perhaps, it also involves an element of art.

2.15 *Economy in blood testing*

I have given you examples which illustrate the triumph of statistics, not so much as data and methodology, the two accepted meanings of statistics, but as a mode of quantitative thinking. I suggest the use of the same word statistics in a third sense to mean quantitative thinking which, when fully codified, will be a fountain source of creativity. I shall give you two more examples.

During the Second World War, a large number of people had to be recruited in the army, and for screening the applicants against certain rare diseases, individual blood tests were suggested which involved a large amount of work. The rejection rate was low but the tests were crucial in determining the fitness of an individual for the army. How does one cut down on the number of tests and yet ensure that the "defectives" are eliminated? There was no textbook answer. Here is a brilliant solution suggested by a statistician.

If only 1 in 20 suffered on the average from a disease, 20 individual tests for each batch of 20 applicants would reveal one positive case (on the average). It is evident that if a number of blood samples are mixed and tested, the mixture will be positive only if one or more individual samples mixed are positive. If instead of 20 individual tests, suppose we make only two tests to begin with, one on the mixed blood samples of the first batch of 10 individuals and another on the mixed blood samples of the second batch of 10 individuals. On the average one mixture will be negative and the other positive. Only in the latter case the individual tests have to be done to find out which of the samples are positive. Thus only $2+10=12$ tests are needed on the average for each batch of 20 samples, which means a reduction in the tests by 8 out of 20, or 40 percent. It may be seen that if mixtures of 5 samples are considered, the total number of tests needed on the average is only $4+5=9$, which is the optimum leading to a saving of 11 tests for each batch of 20 candidates, i.e., 55 percent.

The optimum procedure in similar situations can be found depending on the rate of prevalence of the disease under investigation. Suppose the proportion of affected individuals is π, then the optimum batch size for mixing the samples is the value of n which maximizes the expression $(1-\pi)^n-(1/n)$. For given n, the best way of finding the optimum n is to tabulate the function $(1-\pi)^n-(1/n)$ for different values of n and choose that value for which the value of the function is a maximum.

The idea is beautiful. The procedure can be adopted in other areas. For instance, samples of water from a number of sources are frequently tested for contamination. By adopting the method described of mixing samples and testing in batches, it should be possible to test samples from a larger number of sources and to carry out more elaborate tests on samples without enlarging the resources of a testing laboratory. The method of mixed sample tests is now widely practiced in environmental pollution studies and other areas, resulting in reduced expenditure on testing.

2.16   *Machine building factories to increase food production*

By 1950, India was producing only one million tons of steel

and a proposal was made to build a plant to produce a second million tons of steel. This was however, followed by a survey of current demand for steel by experts, which was estimated to be one and half million tons. On the basis of this figure the wisdom of establishing a new factory for a second million tons was questioned. The proposal was dropped and the alternative of purchasing the extra half a million tons of steel from abroad was recommended.

The decision might have been based on sound economic theory. There seems to be nothing wrong with the arithmetic. But the broader perspective was lost sight of. The problem was not examined in the context of the over-all economic development of the country and the ultimate goal of self sufficiency in different sectors of economic activity. The decision of the expert committee to block the establishment of a new steel plant has cost the country millions of rupees subsequently for the import of steel. Let us see how a statistician looked at the problem (Mahalanobis, 1965).

In India, population is increasing at the rate of 7 million persons per year. The amount of extra foodgrains needed to feed the additional population in the next five years is 15 million tons. If we have to import this, at the world price of about 90 dollars per ton, we would have to spend about 1300 or 1400 million dollars in foreign currency in five years.

To grow 15 million tons of foodgrains we need 7.5 million tons of fertilizers. At a price of 50 dollars per ton the total cost would be less than 400 million dollars in five years. Would it not be wiser to import fertilizers instead of foodgrains.

We can look further ahead. The foreign exchange component in establishing a fertilizer factory is only 50 or 60 million dollars. We may need five such factories to produce the required amount of fertilizers. The total cost would be less than 300 million and there is the additional advantage that the factories will continue to produce fertilizers beyond the five-year period. Would it not be more wiser to set up fertilizer plants instead of importing fertilizers?

We can go a step further and establish a factory to produce

machinery for manufacturing fertilizers, and the cost for this may be only 50 or 60 million dollars in foreign exchange once for all. In this way only 50 or 60 million dollars can serve the same purpose as 300 or 400 or 1400 million dollars. Would it not be still wiser to set up machine building factories?

The argument sounds like the saying: For want of a nail, the horse shoe was lost; for want of a shoe, the horse was lost; for want of a horse, the rider was lost; for want of a rider, the kingdom was lost.

Some of our economists have argued that the Mahalanobian thinking is not in tune with principles of economics; in retrospect we see that Mahalanobis' plan had helped in industrializing India.

2.17   *The missing decimal numbers*

A statistician is often required to work on data collected by others. In many cases the purpose for which the information is collected, sometimes at an enormous cost, is not clearly defined. The first job of a statistician is to interrogate the investigator to understand what his data are about - the population of individuals or objects or locations to which the data refer, the method of sampling employed, the concepts and definitions governing the measurements, the agency employed (individuals and instruments) for obtaining the measurements, the questionnaire used with checks and cross checks if any, whether part of the data is obtained from other sources published or otherwise, and finally what was the object with which the investigation was undertaken and what kind of specific questions are required to be answered on the basis of collected data. There may be communication difficulties between the statistician and the investigator as one may not understand the "language" of the other. This could probably be overcome with a little effort on either side to learn the other's language.

The investigator may be impatient and not appreciate the statistician's desire to understand his problem and the nature of his data, on which solely depend the choice of the statistical techniques to be employed. In such a case he would be behaving like a patient who tells a doctor to prescribe medicine for the ailment he thinks he is suffering from without letting the doctor examine him. It would be unethical for a statistician to accept other's data at face value, put them through the statistical mill and produce end results which may satisfy the customer.

After the dialogue with the investigator, the statistician faces another serious problem. He has masses of data handed over to him - data supposed to be generated according to a particular design chosen by the investigator and recorded without errors. Do the data stand for what they are supposed to be? Can the statistician ascertain this from the given data itself? How does he communicate with figures?

The dialogue between the statistician and the figures, or scrunity of data, is essential and is an exciting part of data analysis. There is no well developed communication channel for this purpose and much depends on the ingenuity of the statistician to make figures speak to him.

In the data given to the statistician, some figures might look suspicious being very low or very high in value compared to others, some might have been recorded without proper identification, and so on. A reference to the original records might be sufficient to resolve some cases. Routine tests of consistency would he of help in few other cases. For the rest, there is no general prescription.

I shall give just one example. A statistician was asked to analyze anthropometric measurements taken on castes and tribes of undivided Bengal. Weight of an individual was one of the ten characteristics measured, and the series of weight measurements (in stones) ran as follows 7.6, 6.5, 8.1, ... The person who edited the measurements converted the values given in stones into pounds by multiplying each figure by 14. Thus the values 7.6, 6.5, 8.1, ..., mentioned to be in stones were expressed in pounds as 14x7.6=106.4, 14x6.5=91.0, 14x8.1=113.4, ... and so on. The statistician, instead of looking at the edited values, thought of going through the original records. He observed what he thought was a peculiarity that in the decimal place of the observations on weight, the digits 7, 8, 9 were completely missing! Something must have happened. The figures as recorded looked innocent, the converted figures appeared plausible and the error would have gone undetected if the original records were not seen. An enquiry revealed that the weighing instrument used in the survey was manufactured in Great Britain and had a dial graduated in stones with 6 marks indicating 7 sub-divisions in between the marks for stones: the investigator was apparently recording the number of stones and the number of sub-divisions shown by the indicator, with a decimal point separating them. The great Hindu invention of the decimal notation was misused! The proper conversion into pounds of the value 7.6 is thus $14x7 + 6x2 = 110$ instead of 106.4 - the loss of 4 to 5 pounds in the average weight of Bengals is thus restored by statistician's alertness (without any nutritional supplement!).

A statistician has to be a detective using his imagination looking for clues and little hints here and there which might unravel a hidden mystery. He should follow the dictum:

> Every number is guilty unless proved innocent.

2.18 *The Rhesus factor: A study in scientific research*

This is a story of how the genetic mechanism of the Rhesus blood group system was uncovered in a short period of time by a group of workers in England. The Rhesus factor was discovered by Levin in 1939 in a case of stillbirth in which the mother's serum was found to contain an antibody referred to as Δ (or anti-D) capable of agglutinating 85% of white American donors. This suggested a possible Mendelian factor with two alleles, the presence of one of which produces the antigen D. Subsequently, to cut the story short, other antibodies were found one after the other, called γ (or anti-c), Γ (or anti-C), H (or anti-E) which produced different combinations of reactions (+ and −) by which at least seven different alleles (or gene complexes) could be distinguished. The reactions of the antibodies γ, Γ, Δ and H with these seven gene complexes designated as R_1, R_2, r, R_0, R'', R', R_z were as shown in the first block of Table 5.5. Judging from the reactions of γ, Γ, Δ, H with the known seven gene complexes, Race (1944) argued as follows and made some predictions.

None of the seven gene complexes react in the same way with respect to γ and Γ, indicating that they are complimentary antibodies. It is quite possible that such complimentary antibodies to Δ and H, to be designated as δ and η respectively, exist.

There is possibly one more gene complex to be designated as R_y with reactions as specified in the last row of Table 5.5 to complete

Table 5.5 Reaction of seven gene complexes to the four
antibodies first known, with extension

gene complex	known antibodies				predicted antibodies		suggested gene complex
	γ	Γ	Δ	H	δ	η	
R_1	−	+	+	−	−	+	CDe
R_2	+	−	+	+	−	−	cDE
r	+	−	−	−	+	+	cde
R_o	+	−	+	−	−	+	cDe
R''	+	−	−	+	+	−	cdE
R'	−	+	−	−	+	+	Cde
R_z	−	+	+	+	−	−	CDE
*R_y	−	+	−	+	+	−	CdE

*Predicted gene complex with indicated reactions.

the system in which each reagent (antibody) reacts positively with four gene complexes, and negatively with four others.

Within a year of these conjectures, Mourant (1945) discovered the antibody η and Diamond the antibody δ.

Fisher (1947) then explained the nature of the gene complexes in terms of three Mendelian factors closely linked with the alleles for each factor designated as (C,c), (D,d) and (E,e). The presence of the genes C, D and E produce positive reactions with the antibodies Γ, Δ and H respectively and the presence of c, d and e produce positive reactions with the antibodies γ, δ and η respectively.

Now we know that the genetic mechanism is more complex with possibly more than two alleles at each of the three loci. However, the investigation involving careful organization of data systematically collected provided a quick and efficient uncovering of what appeared to be a confusing and obscure situation when the Rhesus factor was first discovered.

2.19   *Family size, birth order and I.Q.*

There have been several studies on the decline of average SAT (Scholastic Aptitude Test) scores of high school seniors during the last 20 years. In order to explain this phenomenon, data were collected in a number of countries to study possible association between the children's SAT score, and the parents occupation, family size and birth order. The data on two such studies are given in Tables 5.6 and 5.7.

The data in Tables 5.6 and 5.7 show (except for one figure for family size 1 in Table 5.7) that the scores generally decline with increase in family size and within each family size they decline with birth order (indicating that the later born children are less intelligent than the earlier born).

It is argued that the later born children are brought up under a lower intellectual environment than the earlier born, considering the intellectual environment as the average of the intellectual levels of the

**Table 5.6 Average I.Q. of children in England
classified according to number of sibs in the family**

Number in family	I.Q.	Number of families in sample
1	106.2	115
2	105.4	212
3	102.3	185
4	101.5	152
5	99.6	127
6	96.5	103
7	93.8	88
7+	95.8	102

**Table 5.7 Mean scores on the National Merit
Scholarship Qualification Test, 1965, by place
in family configuration in USA**

Family size	Birth order				
	1	2	3	4	5
1	103.76				
2	106.21	104.44			
3	106.14	102.89	102.71		
4	105.59	103.05	101.30	100.18	
5	104.39	101.71	99.37	97.69	96.87

parents and the earlier born children. A case is made that the effect can be reversed by increasing the age spacing between siblings, so that the intellectual level, depending on age, will be higher for the earlier born at the times of later births.

References

Boneva, L.L. (1971): A new approach to a problem of chronological seriation associated with the works of Plato. In *Mathematics in the Archaeological and Historical Sciences*, Edinburgh University Press, 173-185.

Fisher, R.A. (1938): Presidential Address. First Indian Statistical Conference, Calcutta. *Sankhyā,* **4,** 14-17.

Fisher, R.A., Corbet. A.S. and Williams, C.B. (1943): The relation between the number of species and the number of individuals in a random sample of an animal population. *J. Anim. Ecol.* **12,** 42-58.

 Fisher, R.A. (1947): The Rhesus factor: A study in scientific method. *American Scientist,* **15,** 95-103.

Fisher, R.A. (1952): The expansion of statistics. (Presidential address), *J. Roy. Statist. Soc.* A, **116,** 1-6. *Journal of the Royal Statistical Society*

Halberg, J. (1974): Catfish Anyone? *Chronobiologia,* **1,** 127-129.

Kac, Mark (1983): Marginalia, Statistical odds and ends. *American Scientist,* **71,** 186-187.

Kruskal, J.B., Dyen, I. and Black, P. (1971): The vocabulary method of reconstructing language trees: innovations and large scale applications. In *Mathematics in Archaeological and Historical Sciences*, Edinburgh University Press, 361-380.

Macmurray, J. (1939): *The Boundaries of Science.* Faber and Faber, London.

Mahalanobis, P.C. (1965): Statistics for Economic Development. *Sankhyā* B, **27,** 178-188.

Mosteller, F. and Wallace, D. (1964): *Inference and Disputed Authorship,* Addison-Wesley.

Mourant, A.E. (1945): A New Rhesus Antibody, *Nature,* **155,** 542.

Nita, S.C. (1971): Establishing the linkage of different variants of a Romanian chronicle. In *Mathematics in Archaeological and Historical Sciences*, Edinburgh University Press, 401-414.

Rao, C.R. (1957): Race elements of Bengal: A quantitative study. *Sankhyā* **19,** 96-98.

Race, R.R. (1944): An incomplete Antibody in Human Serum, *Nature,* **153,** 771.

Swadish, M. (1952): Lexico-statistic dating of prehistoric ethnic contacts. *Proc. Amer. Philos. Soc.* **96,** 452-463.

Thisted, Ronald and Efron, Bradley (1987): Did Shakespeare write a newly-discovered poem? *Biometrika,* **74,** 445-455.

Trautmann, T.R. (1971): *Kautilya and the Arthasāstra,* A statistical investigation of the authorship and evolution of the text. E.J. Brill, Leiden.

Yardi, M.R. (1946): A statistical approach to the problem of chronology of Shakespeare's plays. *Sankhyā,* **7,** 263-268.

Chapter 6

Public Understanding of Statistics: Learning from Numbers

Life is the art of drawing sufficient conclusions from insufficient evidence.

Samuel Butler

To understand God's thoughts we must study statistics, for these are the measures his purpose.

Francis Nightingale

1.   Science for all

In his book on *The Social Functions of Science*, published in 1939, J.D. Bernal wrote

> It is no use improving the knowledge that scientists have about each other's work, if we do not at the same time see that a real understanding of science becomes a part of the common life of our times.

It is only half a century later that the importance of what Bernal said is recognized and serious efforts are being made to spread scientific knowledge to the public. National Science Academies of advanced countries have appointed task forces to examine the problem and suggest ways of achieving this. Five years ago, the Royal Society of United Kingdom started a new journal called *Science and Public Affairs* with the broad aim of fostering the understanding of scientific issues by the public and of explaining the implications of discoveries in science and technology in everyday life. The new slogan raised by the Royal Society is

Science is for everybody.

No doubt, science pervades almost everything we do in society, and the importance of public understanding of science needs no emphasis. The public must know how a new technology could be useful to them in improving their standard of life. They must know the consequences of entrepreneurs exploring new discoveries for their benefit disregarding possible harmful effects to the society and to the environment. They should be aware how a government's policy such as establishing nuclear power plants all over the country is going to effect their lives and the lives of their children.

When Bernal wrote the book, statistics was not known as a separate discipline. It grew in importance only in the second quarter of the present century as a method of extracting information from observed data and as the logic of taking decisions under uncertainty. As such, the knowledge of statistics is a valuable asset to people in all walks of life. If Bernal had been alive today to bring out an updated edition of *Social Functions of Science*, he might have added, awed by the ubiquity of statistics, that the public understanding of statistics is far more important than any field of science.

2.   Data, information and knowledge

The only trouble with a sure thing is uncertainty.

What is statistics? Is it science, technology, logic or art? Is it a separate discipline like mathematics, physics, chemistry and biology with a well defined field of study? What phenomena do we study in statistics?

Statistics is a peculiar subject without any subject matter of its own. It seems to exist and thrive by solving problems in other areas. In the words of L.J. Savage

Statistics is basically parasitic: it lives on the work of others. This is not a slight on the subject for it is now recognized that many hosts die but for the parasites they entertain. Some animals could not digest their food. So it is with many fields of human endeavors, they may not die but they would certainly be a lot weaker without statistics.

Statistics has been brought into the academic curriculum in the universities only in the present century. Even now the role of statistics in science and society is not well understood by the public and the professionals.

Not long ago, there were misconceptions and skepticisms about statistics expressed in statements such as the following:

* Lies, damned lies and statistics.
* Statistics is no substitute for judgement.
* I know the answer, give me statistics to substantiate it.
* You can prove anything by statistics.

Statistics was also the subject of jokes such as

* Statistics is like a bikini bathing suit. It reveals the obvious but conceals the vital.

Now statistics has become a magic word to give a semblance of reality to statements we make:

* *Statistics prove* that cigarette smoking is bad.
* *According to statistics,* males who remain unmarried die ten years younger.
* *Statistically speaking* tall parents have tall children.
* *A statistical survey* has revealed that a tablet of aspirin every alternate day reduces the risk of a second heart attack.
* There is *statistical evidence* that the second born child is less intelligent than the first, and the third born child is less intelligent than the second, and so on.

* *Statistics confirm* that an intake of 500 mg of vitamin C every day prolongs life by six years.
* *A statistical survey* has revealed that henpecked husbands have a greater chance of getting a heart attack.
* *A statistical experiment* showed that students do better on a test of reasoning after hearing 10 minutes of Mozart piano sonata than they do after 10 minutes of relaxation tape or of silence.

Statistics as a discipline of study and research has a short history, but as numerical information it has a long antiquity. There are various documents of ancient times containing numerical information about countries (states), their resources and composition of the people. This explains the origin of the word *statistics* as a factual description of a *state*. References to census of people and of agriculture as we know today can be found in the Chinese book *Kuan Tzu* (1000 BC), *Old Testament* (1500) and *Arthasastra of Kautilya* (300 BC).

An example of early record of statistics is figures found on a Royal Mace of an Egyptian king who lived 50 centuries ago (3000 BC). They refer to the capture of

        120,000 prisoners of war
        400,000 oxen
        1,422,000 goats

by the army of the victorious king after a war with another kingdom. How were these nicely rounded figures arrived at? Were they actual counts made by the royal tally keepers or fictitious figures conceived by the active imagination of the victorious king? Was the drastic rounding of figures intended to highlight the large dimensions of the booty? Samuel Johnson believed:

Round numbers are always false.

This must have been anticipated by Weirus, a German physician of the 16th century, a time when most of Europe was gripped by the fear of disease and witches. He calculated that exactly

$$7,405,926$$

ghosts inhabited the earth! Most people believed that the figure must have been the actual count as Weirus was a learned man.

I am reminded of what is recommended in a *Tax Guide* while filing my tax return in the U.S.A.

> Careful scrutiny of GAO reports confirms one important way to reduce the odds of audit. Avoid rounding out dollars when reporting earnings or expenses. Figures of $100, $250, $400, $600 arouse an examiner's suspicion, whereas $171, $313, $496 are less likely to. If you must estimate some expenses, estimate in odd amounts.

The etymological definition of statistics is *data* obtained by some means. What do data convey and how do we use it for a specified purpose? For this, we must know what kind and how much *information* there is in *observed data* for solving a *given problem*. What is information? Perhaps, the most logical definition as given by Claude Shannon, an expert in information theory, is "resolution of uncertainty," which plays a key role in solving a problem. Data by itself is not the answer to a problem. But it is the basic material from which we can judge how well we can answer a problem, how uncertain a particular answer is or what faith we can place in it. The observed data needs to be processed to find out to what extent uncertainty can be resolved. The knowledge of the amount of uncertainty provided by the data is the key to appropriate decision making. It enables us to weigh the consequences of different options and choose one which is least harmful. Statistics as it is understood now is the logic by which we can climb one rung in the ladder from *data* to *information*.

As the information grows gradually reducing uncertainty to a minimal acceptable level, we are several steps up in the ladder in a state of *knowledge* which lends faith to actions we take (subject of course to a small inevitable risk). Such a state of knowledge may not be achievable in all areas and in all situations. This creates the need for statistics as the methodology of taking a decision under the level of uncertainty associated with the given data.

According to the distinguished scientist, Rustrum Roy, knowledge that fits into an accepted body of knowledge and enlarges the scope of knowledge constitutes wisdom, which is one step in the ladder above knowledge. However, it is an old wisdom:

> The road to wisdom?
> Well it is plain and simple to express
> Err
> And err
> Err again
> But less
> AND less
> AND less.

3.   Information revolution and understanding of statistics

A time may not be very remote when it will be understood that for complete initiation as an efficient citizen ..., it is as necessary to be able to compute, to think in terms of averages and maxima and minima, as it is now to be able to read and write.

H.G. Wells

The prosperity of mankind depended in the past on agricultural revolution and later, on industrial revolution. But these have not taken us far in alleviating the misery of people in terms of hunger and disease. The main obstacle to progress has been our inability to foresee the future and make wise policy decisions. Sound policies rest on good information. So there is need to enlarge the data base to reduce uncertainty and make better decisions.

The importance of information as a key ingredient in planning and execution of a project more than expertise in technology is now widely recognized, and we are already witnessing the information revolution as both the public and private enterprises are making heavy investments in acquiring and processing information. It is said that in the USA, about 40 to 50 percent of the employees in public and private sectors are solely engaged in these activities.

That there is demand for statistics by the public is shown by the fact that the newspapers devote considerable space for giving all sorts of information. We have detailed weather prediction for an extended period of about a week to plan our outdoor activities. There are stock market prices to tell us what investments we can profitably make. A special section is devoted to sports to keep us informed of the sporting events from all over the world. A daily newspaper in Edmonton, Canada, publishes what is called daily mosquito index to satisfy the public that the city authorities are doing their best to control the mosquitoes in the city. *New York Times* devotes nearly 30% of its space for all kinds of statistics and reports based on them.

There are magazines like consumer reports informing the public about the prices of commodities and the comparative performances of various products in the market.

There are various levels at which understanding of statistics is important. The first is for the individual, for every one. The need for knowing the three R's (reading, writing and arithmetic) is well known. But these are not sufficient to cope with uncertainties facing an individual at every moment of his life. He has to make decisions while entering college, marrying, making investments and dealing with problems at work every day. This requires a different kind of skill, which we may call the fourth R, statistical reasoning, of understanding uncertainties of nature and human behaviour and minimizing risk in making decisions using his own experience and the collective experience of others. Further, statistical knowledge to an individual will be an asset in protecting himself and his family against infection, guarding himself against propaganda by politicians and unscrupulous advertisements by businessmen, shedding superstition which is worse than disease, taking advantage of weather forecasts, understanding disasters like the radiation leak in nuclear power plants and scores of other things affecting his life on which he has no control.

Does the layman need to make a special study of statistics to acquire the fourth R? The answer is no. A certain amount of statistical education in high school along with arithmetic should be sufficient. Our educational system in schools is more geared to encouraging students to believe in a written word and cautioning them against taking risks symbolized by statements like "Do not count the chickens before they are hatched" instead of preparing them to live in an uncertain world and face situations in the cutting edge of modern life.

We must learn how to take a calculated risk. Recently, there was a press report that among the names carved on Vietnam Veterans Memorial in Washington, there are at least 38 mistakenly listed as dead. When the person responsible for it was asked about it, he said: "I was not positive at the time it was built that the men had been killed because their records were incomplete. I didn't know that it would be possible to add names once the memorial was built. I had the idea these people might be lost to history if we didn't include them."

At the next level, we have politicians and policy makers for whom statistical knowledge is important. The governments have a huge administrative machinery for collecting data. They are meant to be used for making right policy decisions in day-to-day administration and in formulating long range plans for social welfare. The policy makers are expected to seek technical advice in making decisions. However, it is important that they themselves acquire some technical knowledge in understanding and interpreting information. The following anecdotes illustrate the point.

Statisticians working in the Government and in industry are often faced with language barriers with their bosses. The chief of a statistical office, an officer in administrative service, was meeting a group of statisticians who complained that in a report received from another organization some estimates were given without standard errors. [Standard error is a quantitative expression attached to an estimate to convey an idea of the magnitude of error in it.] The chief was reported to have immediately remarked, "Are there standards for errors, too?

A report submitted to the Tea Board by a consulting statistician contained a table with the caption: Estimated number of people taking tea with standard error. Soon a letter was sent to the statistician asking what standard error is, which people take with tea.

A Royal commission reviewing a statistics's report, where it is said that the middle class families have on the average 2.2 children, commented:

The figure of 2.2 children per adult female is in some respects absurd. It is suggested that the middle classes be paid money to increase the average to a rounded and more convenient number.

A health minister was intrigued by the statement in the report submitted by a statistician that 3.2 persons out of 1000 suffering from a disease died during the last year. He asked his private secretary, an administrator, how 3.2 persons can die. The secretary replied,

Sir, when a statistician says 3.2 persons died, he means that 3 persons actually died and 2 are at the point of death.

Government policy decisions are important for they effect millions of people. They need sound information and equally sound methodology for processing information.

Then there are professionals in medicine, economics, science and technology for whom data interpretation and analysis is to some extent a necessary part of their work.

4.   Mournful numbers

Tell me not, in mournful numbers
Life is but an empty dream.

H.W. Longfellow

We are continuously made aware of, through newspapers, magazines and other news media, the good and deleterious effects of our dietary, exercise, smoking and drinking habits, and stress in our profession and other daily activities. The information is given as numbers representing loss or gain in some units. Here are some mournful numbers reproduced from Cohen and Lee (1979).

How do we interpret these figures? What message do they convey? Of what use are they to an individual in shaping his or her life style to enhance happiness? (See Table 6.1.)

Table 6.1 Loss of life expectancy due to various causes.

Cause	Days	Cause	Days
Being unmarried (male)	3500	Alcohol	130
Being left handed	3285	Firearms accidents	11
Being unmarried (female)	1600	Natural radiation	8
Being 30% overweight	1300	Medical x-rays	6
Being 20% overweight	900	Coffee	6
Cigarette smoking (male)	2250	Oral contraceptives	5
Cigarette smoking (female)	800	Diet drinks	2
Cigar smoking	330	Pap test	-4[1]
Pipe smoking	220	Smoke alarm in house	-10
Dangerous jobs, accidents	300	Airbags in cars	-50
Average job, accidents	74	Mobile coronary - care units	-125

[1] Negative number indicates increase in life expectancy

Let us consider the first figure of Table 6.1, which refers to the loss in life expectancy if a male person remains unmarried. This figure can be obtained from the information usually available in death records on sex, marital status and age at death. From the records of males, simply compute the average age at death separately for those married and unmarried. The difference in these averages is the number, 3500 days. This probably provides a broad indication of the hazard due to staying unmarried, speaks good of the institution of marriage and gives a strong case for advising someone to get married fast and *save* about *10 years* of his life! None-the-less, it does not imply a cause [marrying] and effect [living 10 years longer] applicable *to every individual*. It is quite likely that for a specific individual, getting married is suicidal! No doubt, a finer tabulation of the death records by subgroups of males according to various personal characteristics would be more informative. Different groups may have different values for loss or gain in life expectancy. A specific individual may have to analyze his own personality and refer his case to the relevant figure for the subgroup of persons with characteristics similar to his own.

It is seen from Table 6.1 that left-handers die about 9 years younger than the right-hangers. Does this imply that there is something genetically wrong with the left handers? Perhaps not: the difference may be due to the disadvantage the left handers have living in a world where most of the facilities are tailored for use by right handers. However, the statistical information is of some use to a left-hander in protecting himself against possible hazards.

An average, in general, provides a broad indication of a characteristic of a group of individuals (population) as a whole. It serves a useful purpose in comparing populations. Thus we may say that a population of individuals with an average income of $1000 per month is better off than another with $500 per month. An average does not say anything about disparities in the income of individuals. For instance, the individual incomes may vary from $20 to $100,000 and average to $1000. The differences in individual incomes within a population, called *variability*, is also relevant for comparing populations. In most cases, an average and some measure of variability (like the range of incomes) provide information of some practical value. An average by itself may be deceptive and is not, in all cases, useful in making judgements about an individual. Imagine a nonswimmer being advised to cross a river by wading through because his height is more than the average depth of the river!

5.   Weather forecasting

A reliable forecaster is one whose microphone is close enough to the window so that he can decide whether to use official forecast or make up one of his own.

Some years ago weather forecasts used to be in the form of statements like: it will rain tomorrow, it will probably rain tomorrow, no precipitation expected tomorrow, and so on. The forecasts went wrong frequently. But now-a-days weather forecasts read differently: there is 60% chance of precipitation tomorrow. What does 60% mean? Does this statement contain more information than what earlier forecasts implied? Perhaps, to those who do not know what the word "chance" stands for, the present day forecasts may be somewhat confusing and may even give the impression that they are not as precise or not as useful as they used to be.

There is an element of uncertainty in forecasting whatever its basis may be. So, logically speaking, a forecast without any indication of its accuracy is not meaningful or useful for decision making. The quantity such as 60% in the weather forecast provides a measure of accuracy of prediction. It implies that on occasions when such a statement is made, it will rain tomorrow about 60% of the times. Of course, it is not possible to say on which particular occasion it will rain. In this sense, the forecast "there is 60% chance of rain tomorrow" is more informative and a logical one to make instead of issuing a categorical statement like, "it will rain tomorrow." In what sense is this statement useful?

Suppose that you have to decide whether to carry an umbrella or not on the basis of the weather forecast, "there is 60% chance of rain tomorrow." Further suppose that the inconvenience caused to you by carrying an umbrella on any day can be measured in monetary terms as m dollars and the loss to you in getting wet in rain by not carrying an umbrella is r dollars. Then the expected loss in dollars under two possible decisions you can make when the chance of rain is 60% are as follows:

Decision	Expected Loss
Carry an umbrella	m
Do not carry an umbrella	$.6(r) + .4(o) = 6r/10$

You can minimize your loss by deciding to carry an umbrella if $m \leq 6r/10$ and not to carry an umbrella if $m > 6r/10$.

This is a simple illustration of how a measure of accuracy or inaccuracy of prediction can be used to weigh the consequences of different possible decisions and choose the best one. There is no basis for making a decision if the amount of uncertainty in prediction is not specified.

6. Public opinion polls

Once I make up my mind, I'm full of indecision.

Oscar Levant

In the past, kings tried to ascertain public opinion by using a network of spies. Probably, the information so gathered helped them in shaping public policy, enacting laws and enforcing them. The history of modern public opinion polling began with the first publication of Gallup polls. Now public opinion polls have become a routine affair with newspapers and other news media playing a major role in it. They gather information from the public on various social, political and economic issues, and publish summary reports. Such opinion polls serve a good purpose in a democratic political system. They would tell the political leaders and the bureaucracy what the public needs and likes are. They also constitute news informing people on what the general thinking is. This may be of help in crystallizing public opinion on certain key issues.

The results of public opinion polls are usually announced in a particular style which needs an explanation. For instance, the news broadcaster may say:

> The percentage of people who approve the president's foreign policy is 42 with a margin of error of plus or minus 4 points.

Instead of giving a single figure as the answer, he gives an interval $(42-4, 42+4)=(38, 46)$. How is this obtained and how do we interpret it?

Suppose that the *actual* percentage of all adult Americans who approve the president's foreign policy is a certain number, say T. To know the number T, it is necessary to contact all American adults and get their responses to the question: Do you approve the president's foreign policy? This is an impossible task if a timely and quick answer has to be found. The next best thing to do is to get an estimate, which is a good approximation to T. The news media does it by telephoning a certain number of "randomly chosen individuals" and getting their response. If r, out of p persons contacted, respond by saying yes, then the estimate of T is taken as 100 (r/p). Of course, there is some error in the estimate because we have taken only a sample of the people (a small fraction of the number of adults in the USA). If you contact another set of p individuals, you may get a different estimate. How is the error in an estimate specified? Based on a theory developed by two statisticians, J. Neyman and E.S. Pearson, it is possible to calculate a number e such that the true value T lies in the interval

$$100(r/p)-e, \ 100(r/p)+e$$

with a high "chance" usually chosen as 95% (or 99%). What it means is that the event that the interval does not cover the true value is as rare as observing a white ball in a random draw from a bag containing 5 (or 1) white balls and 95 (99) black balls.

The validity of the results of opinion polls depends on "how representative" the choice of individuals is. It is quite clear that the result will depend on the composition of the political affiliations (Republican or Democrat) of the individuals chosen. Even supposing that no bias is introduced in the choice of individuals with respect to their political affiliations, the results can be vitiated if some individuals do not respond and they happen to belong to a particular political party. In any survey, there is bound to be some degree of non-response, and the error due to this is difficult to assess unless some further information is available.

7.   Superstition and psychosomatic processes

When asked why he does not believe in astrology, the logician Raymond Smullyan responds that he is a Gemini, and Gemini never believe in astrology.

A friend of mine, a good christian, donated the whole amount of his first month's salary in his first job to the church. When I asked him whether he believed in God, he replied, "I do not know whether God exists or not, but it would be on the safe side to believe that God exists and act accordingly." Perhaps belief and superstition have a place in one's life, but there is a danger when they become sole guiding factors in one's activities.

Do psychosomatic processes have an effect on biological functioning of the body? There has been no experimental evidence one way of the other. However, some studies are reported from time to tome to support anecdotes concerning the effects of "mind over matter." In a recent study, David Phillips of the University of California, San Diego, examined the death rates over a 25-year period among elderly Chinese-American women around a key holiday, Harvest-moon festival. He found that the deaths dip 31.1% below the norm one week before the holiday and peak 34.6% above the norm a week after, which seems to indicate that one can exercise will power to postpone death until after an auspicious event.

In an earlier study Phillips (1977) obtained data on months of birth and death of 1251 famous Americans and demonstrated similar effects. The following Table 2 gives the data reported by Phillips together with the data on Indian Fellows of the Royal Society.

Table 6.2 Number of deaths before, during and after the birth month

	months before						birth month	months after					Total	p
	6	5	4	3	2	1		1	2	3	4	5		
Sample 1	24	31	20	23	34	16	26	36	37	41	26	34	348	.575
Sample 2	66	69	67	73	67	70	93	82	84	73	87	72	903	.544
Sample 3	0	2	1	9	2	2	3	2	0	1	3	2	18	.611

p = Proportion dying during and after birth month

Sample 1. Very famous people listed in *Four Hundred Notable Americans.*

Sample 2. People mentioned under the category foremost families in 3 volumes of *Who is Who* for the years 1951-60, 1943-50 and 1897-1942.

Sample 3. Diseased Indian Fellows of the Royal Society.

It is seen from Table 6.2 that the numbers of deaths in the months before are smaller than those in the months during and after the birth month. This phenomena is more pronounced in the case of most famous people. The data on the whole seem to indicate that there is a tendency to stave off death until after birthday.

Do these studies indicate that some people can exercise their will power to postpone the date of death till an important event occurs such as a birthday, a festival or an anniversary. A famous example quoted in this connection is that of Thomas Jefferson who is reported to have delayed his death till July 4, 1826 - exactly 50 years after the Declaration of Independence was signed - only after asking his doctor, "Is it the fourth?"

Isolated published studies such as those of David Phillips do not necessarily tell the whole story. In research work, it is not uncommon that the *same problem* is studied by a large number of investigators and only those where positive results occur, perhaps by chance, are reported. Those indicating negative results are not generally reported and remain in filing cabinets, a situation referred to as the "file drawer problem." Therefore, some caution is needed in accepting the results from published sources only and drawing conclusions from them.

8. Statistics and the law

Laws are not generally understood by three sorts of persons, viz. by those that make them, by those that execute them, and by those that suffer if they break them.

Halifax

It is important that not only justice is done but justice is seen to be done.

During the last decade, statistical concepts and methods have played an important role in resolving complex issues involved in civil cases. Typical examples are those of disputed paternity, alleged discrimination against minority groups in employment and housing opportunities, regulation of the environment and safety, and consumer protection against misleading advertisements. In all such cases, the arguments are based on statistical data and their interpretation. A judge has to determine the credibility of the evidence presented to him and decide on the legal liability in each case as well as the appropriate compensation. This process demands that all parties concerned, those involved in dispute, lawyers on either side, and, perhaps more importantly, judges who decide, have some understanding of statistics and the common pitfalls in the use of statistics.

Let us consider the case of *Eison versus City of Knoxville*, in which a female candidate at the Knoxville Police Academy claimed that a test of strength and endurance used by the Academy discriminates against the female sex. As evidence, she produced the test results in her class.

Table 6.3 Pass rates for persons in plaintiff's class

Sex	Pass	Fail	Percentage passing
Female	6	3	.666
Male	34	3	.919
Total	40	6	.870

She said the four fifths rule of EEOC (Equal Employment Opportunity Commission) is violated since the ratio .666/.919 = .725 is much less than (4/5) = .8. The judge asked for the results of the Academy as a whole which were as given in Table 6.4. In this case, the ratio (.842)/(.955) = .882 > .8. The judge quite rightly said that what is relevant is the "universe of persons" taking the test and not a particular "subset." This is a typical example where interested parties try to choose subsets of data which seem to differ from the entire body of data and make a case.

Often, the quantitative evidence produced is in the form of an average or a proportion, based on a survey of a small proportion of individuals of a population, for a particular measurement or opinion.

Table 6.4 Pass rates for all persons in the academy

Sex	Pass	Fail	Percentage passing
Female	16	3	.842
Male	64	3	.955
Total	80	6	.930

Does the quoted figure represent the particular characteristics of the population as a whole? Much depends on the adequacy of the number of individuals contacted and the absence of bias in their selection.

The acceptance of sample estimates for population values requires a careful examination of the processes followed in conducting the survey, such as ensuring the representativeness of the sample and using adequate sample size to ensure a certain degree of accuracy of estimates. Justice would be better served if the judges have some understanding of the survey methodology to enable them to decide in individual cases whether to accept or reject sample estimates. It is not suggested that a judge has to be a qualified statistician, but some exposure to statistical inference and the uncertainties involved in decision making will be an asset to a judge in forming an independent opinion on statistical arguments presented to him.

Any judgement involves the evaluation of the degree of proof or probability that an event is true given all the evidence and taking a decision considering the consequences of convicting an innocent person and failing to convict a guilty one. The standards for various degrees of proof are expressed in verbal terms such as:

(1)　the preponderance of the evidence;
(2)　clear and convincing evidence;
(3)　clear, unequivocal and convincing evidence;
(4)　proof beyond reasonable doubt.

In order to ascertain how judges generally interpret these standards of proof, judge Weinstein surveyed his fellow district court judges, whose probabilities expressed as percentages are given in Table 6.5.

It is seen that there is consistency in the increasing order of probabilities assigned by the judges for the four standards listed above. However, there is some variation between judges in the probabilities assigned to higher order degree of proof.

Indeed, there exists a sophisticated statistical technique in statistics, the Bayes procedure by which a judge's *prior probability* that an individual is guilty can be updated by using current evidence of a given degree of credibility. This probability conditioned on

Table 6.5 Probabilities associated with the various standards of proof by the judges in the eastern district of New York

Judge	Preponderance (%)	Clear and convincing (%)	Clear, unequivocal and convincing (%)	Beyond a reasonable doubt (%)
1	50+	60-70	65-75	80
2	50+	67	70	76
3	50+	60	70	85
4	41	65	67	90
5	50+	Standard is elusive and unhelpful		90
6	50+	70+	70+	85
7	50+	70+	80+	95
8	50.1	75	75	85
9	50+	60	90	85
10	51	Cannot estimate numerically		

Source: U.S.v.Fatico 458 F. Supp.388 (1978) at 410.

current evidence is called the *posterior probability* which is the main input in decision making. It appears that the theory of Bayesian decision making as developed in statistics provides an objective basis for administering justice.

9.   ESP and amazing coincidences

> *The universe is governed by statistical probability rather than logic. But that still makes it wonderful. If life is like throwing a six hundred times in succession, we know that it is not likely to happen oftener than once in so many centuries, but we also know it could happen in this room tonight without upsetting the cosmic apple cart. This is reassuring.*
>
> G.K. Chesterton

From time to time we come across reports about individuals possessing extra sensory perception (ESP) with the ability to read the minds of others, astrologers making accurate predictions and amazing coincidences like someone winning a lottery twice in four months. Such events do make news and perhaps they are interesting to read. Do they suggest the existence of hidden powers causing them?

It is perhaps, not prudent to completely rule out the possibility that individuals with extraordinary abilities (like ESP) exist and that the positions of the planets at the time of birth determine the course of events in an individual's life. However, reporting of success stories, often on a selective basis, do not provide strong evidence of such possibilities.

Consider, for instance, a typical ESP experiment where a subject is asked to guess which one of two possible objects the experimenter has chosen and places it under a cardboard. The chance of an individual coming up with all correct answers in four repeated trials by pure guessing is 1/16. This means that if 64 individuals from a general population are tested, there is a high chance of *some* 3 or

4 individuals giving all correct answers by pure chance. Such an experiment does not suggest that *these* 3 or 4 individuals have ESP. However, if only their performances gets reported, it would attract our attention.

Let us consider another example. If you are in a party with at least 23 people and ask them to give you their birthdays, you may find that two of them have the same birthday. This may appear to be an amazing coincidence, but probability calculations show that such an event can occur with a 50% chance.

In a paper published in the Journal of the American Statistical Association (Vol. 84, pp.853-880), two Harvard professors, Diaconis and Mosteller, show that most of the coincidences, such as someone somewhere in the US winning a lottery twice in four months, which may appear as amazing, are events which have a fair probability of occurrence over a period of time.

There is a law in statistics which states that with a large enough sample any event, however small its chance may be in a single trial, is bound to occur. It may occur anytime and no special cause can be attributed to it.

10.   Spreading statistical numeracy

I wish he would explain his explanation.
Lord Byron

We learn the 3 R's in the school - reading, writing and arithmetic. These are not sufficient. There is a greater need to know how to handle uncertain situations. How do we take a decision when there is insufficient information? Attempts should be made to introduce the fourth R, reasoning under uncertainty, in the school curriculum at an early stage. This can be done by giving examples of unpredictable events in nature, variability among individuals and errors of measurements, and explaining what can be learnt from observed data or information in such situations.

We should also explore the possibility of using the news media, newspapers, radio and television for continually educating the public on the consequences of actions taken by the government and the findings of the scientists. This needs knowledgeable reporters with the ability to interpret statistical information and report on them in an unbiased way. No doubt, news reporters have some limitations. They have to write stories in such a way that they do not offend the establishment and are sensational enough for acceptance by the editors for publication. They may not have the expertise for independent judgement and prefer to summarize what the experts want to promote. Perhaps, there is a need to train reporters for reporting on statistical matters. I understand that Professor F. Mosteller of Harvard University gives periodical courses on statistics to science reporters to enable them to write about statistical matters unbiasedly and in a way intelligible to the public. This is a worthy attempt and efforts should be made to introduce regular courses for science writers in the universities.

11.   Statistics as a key technology

In the past, the economy of a country depended on how well it was preparing for war. We are witnessing today a transformation from threats and confrontation to conciliation and negotiation. The biggest problem of the coming decades for any country is not the challenge of war but of peace. The battle ground of the future is going to be economic and social welfare where we have to fight hunger and deprivation afflicting the society. We do not seem to be fully prepared for the attack. Our success will depend on acquiring and processing the information needed for optimum decision making by which the available resources, both in men and material, are put to maximum use for improving the quality of life of individuals. This has to be done in a careful way to ensure the following:

*   The progress is equitable and sustainable.

* No irreversible damage is done to the biosphere.

* There is no moral pollution (or degradation of human values).

In achieving this revolution, statistics would be the key technology. a technology for shaping a new world through peace.

References

Cohen, B. and Lee, I.S. (1979). A catalog of risks, *Health Physics*, 36, 707-722.

Diaconis, P. and Mosteller, F. (1989). Methods for studying coincidences. *J. Amer. Statist. Assoc.*, **84**, 853-880.

Phillips, D.P. (1977). Deathday and birthday: An unexpected connection. In *Statistics: A Guide to Biological and Health Sciences* (Eds. J.M. Tanur, et. al.), pp.111-125, Holden Day Inc., San Francisco.

Appendix: Srinivasa Ramanujan
- a rare phenomenon

I consider it a great honor to be called upon to deliver the CSIR Ramanujan Memorial lectures. I have accepted this assignment with great pleasure, especially because Ramanujan's life has been a great source of inspiration to the students of my generation. The birth centenary of this great genius we are celebrating this year is significant in many ways. It reminds us that the mathematical tradition in India, which began with the fundamental discoveries of zero and negative numbers still exists. It will be a reminder to the younger generation that they too can enrich their lives through creative thinking. Finally, I hope it will generate national awareness of the importance of mathematics as a key ingredient of progress in science and arts, and remind us that all efforts should be made to encourage the study of and research in mathematics in our country.

In 1986, the President of the United States of America proclaimed the week of April 14 through April 20 as National Mathematics Awareness Week to keep up the interests of American students in studying mathematics. The spirit of the Soviet Sputnik still haunts the United States and the tendency to neglect mathematics is looked upon as a setback to the scientific and technological advancement of the country. More than a proclamation of National Mathematics Awareness Week, what we need in India is a declaration of our lack of awareness how weak we are in mathematics. Let us dedicate the birth centenary year of Ramanujan to the advancement of mathematics in India. Let it not be said that out contribution to mathematics began with zero and ended there.

I would like to say a few words about Srinivasa Ramanujan as his life and work has something to do with the topic of my lectures. Ramanujan appeared like a meteor in the mathematical firmament, rushed through a short span of life and disappeared with equal suddenness at the age of 32. In the process, he put India on the map of modern mathematics. Ramanujan's mathematical contributions in many fields are profound and abiding, and he is ranked as one of the world's greatest mathematicians. Ramanujan did not do mathematics as mathematicians do. He discovered and created mathematics. This makes him a phenomenon and an enigma, and his creative process a myth and a mystery.

At the time of his death, he left a strange and rare legacy: about 4000 formulae written on the pages of three notebooks and some scraps of papers. Assuming that the bulk of his work was produced during a period of 12 years, Ramanujan was discovering one new formula or one new theorem a day, which beats the record of anyone involved even in a less creative activity. These are not ordinary theorems; each one of them has the nucleus of generating a whole new theory. These are not a number of isolated magical-seeming formulae pulled out of thin air, but something which have profound influence on current mathematical research itself and also in developing new concepts in theoretical physics from the superstring theory of cosmology to statistical mechanics of complicated molecular systems. The work of his last one year of life, while his health was decaying, recorded by hand on 130 unlabelled pages, was discovered in 1976 in the library of Trinity College, Cambridge. The results given in his "Lost Notebook" alone are considered to be "equivalent of a lifetime work for a great mathematician." Commenting on the originality, depth and permanence of Ramanujan's contributions, Professor Askey of the University of Wisconsin said:

> Little of his work seems predictable at first-glance, and after we understand it, there is still a large body of work about which it is safer to predict that it would not be rediscovered by any one who has lived in this century. Then there are some of the formulae Ramanujan found that no one can understand or prove. We will probably never understand how Ramanujan found them.

It is difficult to understand Ramanujan's creativity; there is no parallel in the annals of scientific research or fine arts. Ramanujan discovered the mysterious laws and relationships that govern the endless set of integers just as a scientist tries to discover the hidden laws governing natural events in the universe, but in a style that awes and frustrates any scientist. Let us look at Ramanujan's conjecture in 1919, shortly before his death, about the function $p(n)$, defined combinatorially as the number of distinct ways of expressing an integer as the sum of integral parts ignoring the order of the parts:

"If $24n - 1 \equiv 0 \mod (5^a 7^b 11^c)$,

$$(1)$$

then $p(n) \equiv 0 \mod (5^a 7^b 11^c)$."

The idea behind the formula is superb and the form of the result is a beautiful discovery as nothing of this kind was available in the general theory of elliptic functions or modular functions over a century before. It was shown by another Indian mathematician Chowla that the conjecture is wrong as it does not hold for $n = 243$. The formula needed only a slight modification:

"If $24n - 1 \equiv 0 \mod (5^a 7^b 11^c)$,

$$(2)$$

then $p(n) \equiv 0 \mod (5^a 7^{(b/2)+1} 11^c)$"

with b in the exponent of 7 in the second line of (1) replaced by $(b/2) + 1$, as shown by Atkin (1967), [*Glasgow Math. J.*, Vol, 8, pp. 14-32]. That Ramanujan did not obtain the correct formula, which he might have if he employed mathematical reasoning, is relatively unimportant; that he conceived the idea of such structural property demonstrates unexplainable thought processes behind its discovery.

How does one get a brilliant idea? What kind of preparation is needed for the mind to become creative? Is a genius born or made? There may not be definite answers to these questions. However, even if answers could be found, we may not be able to explain the rapidity with which brilliant ideas emanated from Ramanujan's brain. It is all the more intriguing since Ramanujan had no formal education in higher mathematics, was never initiated into mathematical research and was unaware of the problem areas or trends of research in modern mathematics. He stated theorems without proofs, and without indicating what the motivation was. Ramanujan could not explain how he obtained the results. He used to say that the goddess of *Namakkal* inspired him with the formulae in dreams. Frequently, on rising from bed, he would note down some results and rapidly verify them, though he was not always able to supply a rigorous proof. Many of Ramanujan's stated theorems are proved to be correct. Does creativity take place at the subconscious level?

Professor P.C. Mahalanobis was a contemporary of Ramanujan at Cambridge. He used to narrate several incidents connected with Ramanujan, which are recorded in the biography, *Ramanujan, the Man and the Mathematician* by S.E. Ranganathan. I shall quote from Ranganathan's book one of the incidents as recollected by Professor Mahalanobis.

> On one occasion, I went to his (Ramanujan's) room. The first world war has started sometime earlier. I had in my hand a copy of the monthly *Strand Magazine* which at that time used to publish a number of puzzles to be solved by readers. Ramanujan was stirring something in a pan over the fire for our lunch. I was sitting near a table, turning over the pages of the magazine. I got interested in a problem involving a relation between two numbers. I have forgotten the details; but I remember the type of the problem. Two British officers living in two different houses in a long street have been killed in the war; the door numbers of these houses were related in a special way; The problem was to find these numbers. It was not at all difficult. I got the solution in a few minutes by trial and error.
>
> I said (in a joking way): Now here is a problem for you.
> Ramanujan: What problem, tell me. (He went on stirring the pan.)
> I read out the question from the *Strand Magazine*.
> Ramanujan: Please take down the solution. (He dictated a continued fraction.)

The first term was the solution which I obtained. Each successive term represented successive solutions for the same type of relation between two numbers, as the number of houses in the street would increase indefinitely. I was amazed. I asked: Did you get the solution in a flash?

Ramanujan: Immediately I heard the problem, it was clear that the solution was obviously a continued fraction; I then thought, "which continued fraction?" and the answer came to my mind. It was just as simple as this.

According to Ranganathan, the first occasion when Ramanujan was known to have shown interest in Mathematics was when he was 12 years old. He was then said to have asked a friend studying in a higher class of the Town High School in Kumbakonam about the "highest truth" in mathematics. The theorem of Pythagoras and the problem of Stocks and Shares were said to have been mentioned to him as the "highest truth"! Pythagoras theorem belongs to proper mathematics where conclusion are drawn from given premises through a series of deductive arguments and there is no question of uncertainty about the conclusions. The problems of stocks and shares belong to probability, where the conclusions drawn are not necessarily correct, but helpful to the speculator. Both are intellectually challenging areas of study and research and it is perhaps familiarity with Pythagoras theorem rather than with stocks and shares that might have led to Ramanujan's involvement with mathematics.

Ramanujan recorded most of his results in notebooks without proofs. It is said that he did all his derivations on a slate using a slate pencil and recorded only the final result on paper. When asked as to why he was not using paper, he said that would consume three reams of paper per week and he did not have the money for that.

Ramanujan had 5 papers published in Indian journals before he went to Cambridge in 1914 to work with the world famous Cambridge Mathematician, G.H. Hardy. There are altogether 37 published papers written by himself or jointly with G.H. Hardy distributed over the short span of his working years as follows.

Period	-1914	1914	1915	1916	1917	1918	1919	1920	1921
Number of papers	5	1	9	3	7	4	4	3	1

Ramanujan died in 1920 at the age of 33. The last two to three years of his life was the period of his declining health, during which he continued to work and left behind numerous results recorded in a notebook, which was discovered a few years ago. This "Lost Notebook" has a number of new theorems which have opened up new areas of research in number theory.

Of course, Ramanujan was a rare phenomenon and he blossomed in a more or less hostile environment in which he lived - a routine educational system geared to produce clerical staff for administrative work, poverty which forced brilliant students to give up academic pursuits and take up employment for living and lack of institutional support or other opportunities for research. Referring to Ramanujan's achievements in mathematics, Jawaharlal Nehru wrote in his *Discovery of India*:

> Ramanujan's brief life and death are symbolic of conditions of India. Of millions, how many get education at all? How many live on the verge of starvation? If life opened its gates to them and offered them food and healthy conditions of living and education and opportunities of growth, how many among these millions would be eminent scientists, educationists, technicians, industrialists, writers and artisans helping to build a new India and a new world?

Jawaharlal Nehru was a visionary. The conditions in India seem to be very much improved over the years and the average level of science in India now is, indeed, comparable to that of any advanced country. But there is a general feeling that we have not reached the desired level of excellence. I hope our government and academic bodies will investigate (with the help of statisticians!) and do what needs to be done to place India in the forefront of innovation and scientific sophistication.

Index